Fundamentals of Radiation and Chemical Safety

Fundamentals of Radiation and Chemical Safety

Ilya Obodovskiy

ELSEVIER

AMSTERDAM • BOSTON • HEIDELBERG • LONDON
NEW YORK • OXFORD • PARIS • SAN DIEGO
SAN FRANCISCO • SINGAPORE • SYDNEY • TOKYO

Acquiring Editor: Kostas KI Marinakis
Editorial Production Manager: Sarah Jane Watson
Project Manager: Surya Narayanan Jayachandran
Designer: Mark Rogers

Elsevier
Radarweg 29, PO Box 211, 1000 AE Amsterdam, The Netherlands
The Boulevard, Langford Lane, Kidlington, Oxford OX5 1GB, UK
225 Wyman Street, Waltham, MA 02451, USA

British Library Cataloguing-in-Publication Data
A catalogue record for this book is available from the British Library

Library of Congress Cataloging-in-Publication Data
A catalog record for this book is available from the Library of Congress

ISBN: 978-0-12-802026-5

For information on all Elsevier publications
visit our website at http://store.elsevier.com/

Table of Contents

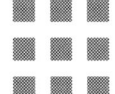

Introduction

The nature and human society may offer us a wide variety of hazards. But, in response, the civilization has formulated simple rules that could make it possible to exclude much of the life risks or, at least, to minimize the unpleasant consequences. One has to wash hands before eating, buckle their seat belts, and don't drink and drive. It is necessary to observe rules of the road that concern both, the driver and the pedestrian. One must also keep in mind that some of the mushrooms are rather poisonous.

We live in the era of technological revolution, where science and technology are becoming more and more influential in our lives. The ionizing radiation was discovered more than 100 years ago. Now, it is hard to imagine our life without X-rays, radioisotope diagnosis and therapy, and the wide range of radiation methods of analysis and measurement [1]. The significant role in meeting the energy needs of humanity plays the nuclear power. There is no doubt that after a period of retreat and improvements, the expansion of its use would continue.

Ionizing radiation has always been present in our lives even before it was discovered, and judged by the experimental studies mentioned in this book, it has played and still continues to play an important positive role in human lives, in spite of the fact that the radiation intensities, used in practice, have extremely increased.

Since the discovery of radiation, it became clear that ionizing radiation can be a certain harm or a great benefit to the people. For useful applications, see, for example, [1]. The psychological stress one way or other developing in human society is the evilest effect of radiation. But the loss of human lives and the property damage resulting from careless use of ionizing radiation and nuclear energy, in general, has been for a 100 years of use of nuclear technology significantly less than damage of fire, of road accidents, of aviation, railway, and water transport accidents, and also of different chemical invasions (including bombings) upon the human environment.

The high reliability of the work with ionizing radiation, which, at most, is able to destroy all the life on earth, is largely determined by the fact that the radiation detection is highly reliable and secure. The effects of radiation on living organisms and on the various materials used by man in his practice have been studied in detail. The rules for handling radiation sources are clearly identified and formulated, and if these rules ever strictly followed, the work with ionizing radiation could be one of the safest areas of human activity. In all the problems associated with the threat of radioactive destruction, the so-called human factor deserves the most attention. It's not the radiation itself that is to be blamed for the possible disaster. The main point here is how real people can handle it.

After a rather long existence in a state of alchemy, the chemistry as the real science has created preconditions for the development and a wide application of various synthetic substances in industry, agriculture, and everyday life. The rapid development of science and technology has given humanity new and outstanding possibilities and alongside with it has led to some unwanted side effects with their detrimental consequences for human health and environment. Quite lot new and unhealthy chemicals for humans, the so-called xenobiotics that do not belong to the natural biotical circulation, appeared in the environment.

To analyze the impact of chemical substances on human health so as to prevent the ill effects of dangerous stuff is not at all easy because of certain difficulties to detect them and reveal their real danger that is more evil than the impacts of the ionizing radiation.

Nevertheless, traditionally and psychologically, the human society has developed some strong preconception against ionizing radiation along with the much more loyal perception of the dangers connected with the impacts of chemical substances.

At present, there have been published quite a lot of scientific studies and textbooks on safety of the vital activity. A lot of references and links one can find on the site of National Fire Protection Association [2]. A large number of materials are available on the Russian website [3]. The main subject is, as usual, the safety in case of emergency, such as natural or technological disasters.

This book mainly focuses on the impacts of small doses of radiation as well as of chemical substances, and there are at least two main reasons for that.

First reason: the small radiation and chemical doses are the threat both to the humanity at whole and to its individuals, irrespective of their wish and their way of life, and not only during social conflicts, wars, and natural disasters. Such impacts cannot be easily controlled and practically impossible to avoid. It all concerns with natural radiation background together with a radiation that is considered as a side and, in general, unwanted product of human activity, as well with a radiation that has humane functions of radiation diagnostics and radiation therapy being in itself potentially dangerous nuclear radiation. It also concerns with permanent effect of various chemical substances in the air we breathe, in liquids we drink, in our food, in our medicals, in means of personal hygiene, in mineral fertilizers and pesticides, in food products, or in industrial wastes.

Second reason: the study of effect of the smaller doses upon human body brings up the most important scientific and practical questions of impact mechanisms, of existence of threshold effects and the area of hormesis. It is the most interesting scientific problem today.

We think that the hazards caused by ionizing radiation and chemicals could and should be simultaneously considered. There is much in common in the impact mechanism of the both affecting factors. Also many similarities are in the research methods as well as in the ways of data processing. And, at last, but not the least, the final impacts could in many ways be congruent.

Everyone nowadays should be aware of the real sources of danger as well as of the rules of personal respond to them. This is the necessary point in modern civilization and culture. The more of concern it is for the scientists. Lots of doubtful and often defective information can be found on pages of various papers and on television.

The author expects that this book would help the reader to get the necessary information in order to have the right orientation in discovering the real sources of danger.

In this book, the reader is being introduced with the two different fields of knowledge – nuclear physics and biology – with the structure of the nucleus as well as with the constitution of the cells. The scientists, who have to deal with the both, have concluded that a cell is more complicate than a nucleus. One of the major arguments for this refers to the mathematics. A great bulk of nuclear physics can be determined by calculation, that's why there is so much mathematics in nuclear physics. In molecular biology, the mathematics plays much lesser role. That comes because of insufficient knowledge and comprehension of the processes in the cells, and, so far, still takes place the further complication of our knowledge about the functioning of the cell and the whole human body.

The optimum for a physicist is to be able to interpret a complex phenomenon using a minimum of characteristics. For example, the Ohm's law interprets the movement of electrons in the substance. The whole process is extremely complicated. The electrons move through potential barriers and potential wells, collide with the heterogeneity of substance, change their route, give and take energy, and so on. The complete account of electron movement acquires a great number of parameters: the power characteristics of potential wells and barriers, the probability of collisions, the concentrations, and so on. But as it is, one could forget about all those complications by bringing in one single parameter, the resistance.

In biology, there also exists the equal conception of expediency of such reduction. The central dogma of molecular biology has played here a real positive role. The leading American biologists D. Hanahan and R. Weinberg wrote in an article for the journal *Cell* [4] that "We foresee cancer research developing into a logical science, where the complexities of the disease, described in the laboratory and clinic, will become understandable in terms of a small number of underlying principles." But, so far, biology to a great extent remains a descriptive science. Quite probable is that the abundance of mathematics in this book could make certain difficulties for the students of related sciences as well as for the other reading public.

In the Preface to his book "The Brief History of Time," Stephen Hawking wrote [5] that he was warned that for every equation in the book, the readership would be halved. He meant, though, a popular science book or more exactly "buyers in airport bookshops that Hawking wished to reach" [6]. Modern science is absolutely unfeasible without mathematics.

The issues in this book could be referred to the sphere of interdisciplinary sciences. This assumes that the experts in one field, say, biology, have to comprehend the scientific language of those who deal with physics, and vice versa. It is commonly known that every science has its own complicated vocabulary that could remain obscure for the experts in other sciences. As the Russian biologist M. Frank-Kamenetzky once wrote in the story about the physicist Max Delbruck, who at that time began to take interest in biology (and then became a famous Nobel-prize 1969 winner biologist) [7]: "... this devil vocabulary, as if purposely invented to scare away the uninitiated. When he (Max Delbruck) happened before to watch the genetics speaking, he wondered, why did they have to devise that specific cryptographic language. What if they are a gang of robbers? After all only the criminals invent their peculiar jargon in order to hide from the public their criminal intentions." There is a well-known witticism to illustrate the special character of one of the sciences: "The recessive allele does not impact on the phenotype unless the genotype is homozygous."

Still there is no other way out for the scientists as to learn to effectively communicate with each other. The author was trying to do his best in order to settle the problem somehow, use the most comprehensive vocabulary, or interpret them on the spot.

Summing up what have been said before, we live in the era of a civilizational revolution that could be compared in its outcome with discovery and implementation of printing. E-books are taking place of traditional paper books. It is particularly important for all kinds of academic literature. Up to date, there have been accumulated in the Internet quite a lot of interesting and quite solid stuff for those who would want to broaden their knowledge on the subject–matter of this book. The e-stuff has a sound advantage over the traditional paper books and magazine articles because of:

their easy availability, easy storage, and use;
practically unlimited scope;
easy availability of color pictures and figures, animations, and video materials;
easy availability of references to certain items and terminology, notes and comments, various books, cross-references, and so on;
easy availability of various materials from the Internet storage;
opportunity to publish the stuff gradually, making necessary additions or expanding the whole thing.

But the e-academic stuff is not quite free from certain drawbacks, such as:

occasional uncertainty of dating;
occasional uncertainty of authorship;
possible disappearance of the resource;
certain problems with the copyright.

The author is making many references to electronic resources. Each time the cited articles and books here, if they are available electronically, are supplied with the website address, they could be taken either without any limitation or by the necessary registration.

The electronic resources are, no doubt, going to be more affluent and the access to them is going to be more simplified with the drawbacks by-and-by decreasing.

This book is oriented to the students, postgraduates, and researchers in physics, chemistry, biology, ecology, and also to those of the wide range of interdisciplinary science. It can be also useful as a reference manual for those who are in search of the sources for more detailed information on special interests on radiation and chemical safety. Those who are interested with the problems of life safety could also find here much useful information. This book could be of interest as well for a wide range of those inquisitive readers, who may take a keen interest in real sources of danger within the everyday life.

The author has made use of certain data, tables, and plates from different printed sources and online materials converting them, as a rule, for the purposes of his book. Every time the source is tagged by the link.

This book has been written and primarily published in Russian. The translation has been done in part by the author, and in part by Oxana Kirichenko and Elena Evseeva. The author is grateful for both of them. For the translated edition, the author excluded those references to Russian papers and links to sites in Russian that, to his mind, are not interesting for international readers. Instead, he included some new references and links to materials in English.

The author expresses his gratitude to his colleagues, whose professional and friendly advices and suggestions helped to significantly improve the whole content of the book and the manner of its presentation. They are Professor G. Belitsky, PhD, Institute of Carcinogenesis, N. N. Blokhin, Cancer Research Centre Russian Academy of Medical Sciences, Moscow, Russia; G. Bakale, PhD, Case Western Reserve University, Cleveland, OH; K. Kitchin, Health Effects Research Laboratory, US Environmental Protection Agency, Triangle Park, NC; Professor A. G. Khrapak, PhD, United Institute for High Temperature, Russian Academy of Sciences, Moscow, Russia; S. Pokachalov, PhD, National Research Nuclear University MEPhI, Moscow, Russia; and I. Kandror, PhD, independent researcher, Wiesbaden, Germany. The author would also like to give special thanks to Oxana Kirichenko for her everyday encouragement and patience.

References

[1] Bolozdynya A, Obodovskiy I. Detectors of ionizing particles and radiations. Principles and applications. Dolgoprudny: Intellect; 2012, p. 204 (in Russian: А.И. Болоздыня, И.М. Ободовский. Детекторы ионизирующих частиц и излучений. Принципы и применения. Долгопрудный: Изд. Дом Интеллект, 2012, 204 с).

[2] National Fire Protection Association. http://www.nfpa.org; 2014.

[3] List of literature on emergency. http://www.library.ugatu.ac.ru/pdf/diplom/ch_s.pdf; 2013.

[4] Hanahan D, Weinberg R. The hallmarks of cancer. Cell 100; 2000, p. 57–70.

[5] Hawking S. A brief history of time. Westminster, Maryland: Bantam Dell Publishing Groups; 1988, p. 256.

[6] http://en.wikipedia.org/wiki/A_Brief_History_of_Time.

[7] Frank-Kamenetzky M. The main molecule (in Russian). Moscow: Nauka, library Quantum; 1983, p. 159.

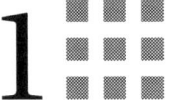

1 Basics of Nuclear Physics

1.1 Peculiarities of the Processes in Microcosm

The processes in which ionizing radiations participate, as well as their production and interaction with matter, are taking place in the microcosm. It is reasonable to recall what are the peculiarities of the microcosm and how they differ from our familiar world of macroscopic bodies.

1. Microparticles exhibit not only particle properties but also wave properties. Particles can be described not only by the pulse p and the laws of movement of material bodies but by the wave length λ. The connection of these values is given by the expression of the de Broglie wave $\lambda = h/p$, where h is the Planck constant. The dimension of the Planck constant is the product of energy and time. In mechanics, such value is called action. In many problems of quantum mechanics, in the case of spherical geometry, the value $h/2\pi = \hbar$ appears. Correspondingly, in the case of the de Broglie wave, the reduced value $\bar{\lambda} = \hbar/p$ is used.

   In the case of a wave form of movement, the principle of superposition plays an important role. It means that waves in a "collision" overlap, increasing or decreasing their amplitude depending on the phases, and then they diverge unchanged, unlike particles that in a collision change their energy and the direction of movement.
2. Electromagnetic radiation reveals not only the wave properties but also the corpuscular ones. The energy of electromagnetic quantum (photon) with frequency ν is expressed by the Planck formula $E = h\nu$. Quantum of electromagnetic radiation has no rest mass, but has a relativistic mass $m = h\nu/c^2$ and a momentum $p = h\nu/c$, where c is the speed of light.

   At the beginning, L. de Broglie, who was the first to suggest the idea of wave properties of matter, and then other physicists, supposed that wave property is a fundamental property of matter, or that matter is "spread" over space. Correspondingly, a theory of particle behavior in the microcosm that soon appeared received the name "wave mechanics." But then, mainly due to work by M. Born, it became evident that de Broglie waves are not the real waves of matter but they only reveal the opportunity to reveal microparticles. So, electron wave function characterizes the probability of electron finding, but electron itself is a point particle.
3. Particles in microcosm, if in bonded state, can take only definite energetic levels; that is, their states are quantized. Angular momentum and some other characteristics are also quantized. As quantization is one of the most important properties of microcosm, and "waves of matter" are virtual ones, then the name "wave mechanics" has been changed for "quantum mechanics."

Although processes in the microcosm are ruled by quantum mechanics, the quantum-mechanical concept and methods of calculation should not be necessarily used in all cases. Common experience, elementary considerations of common sense, and traditions of academic physics lead to the fact that principles of material body behavior and the principles of classical mechanics are more evident and are understood more easily and deeply than the principles of wave behavior and the concepts of quantum mechanics. The classical description is simpler and clearer than the quantum one, so it is used in all cases when accuracy of results obtained is sufficient.

4. One of the fundamental principles of quantum mechanics is the uncertainty principle – which states that the energy and pulse of a particle or the energy and duration of a process cannot be noted simultaneously with an unlimited degree of accuracy. The relations that express the uncertainty principle are as follows

$$\Delta p \, \Delta x \geq h; \tag{4.1}$$

$$\Delta E \, \Delta t \geq h; \tag{4.2}$$

where Δp, Δx, ΔE, and Δt are the uncertainties in the value of pulse, coordinate, energy, and process duration, respectively.

5. The particles in the microcosm have their own mechanical moment, called "spin." Spin is a solely quantum-mechanical phenomenon; it does not have a counterpart in classical mechanics (despite that the term *spin* is reminiscent of classical phenomena such as a planet spinning on its axis). Spin is a vector quantity; it has a definite magnitude and a direction. The spin value is measured in fractions of $\hbar$. Depending on the type of a particle, it can be equal to 0, $\hbar/2$, $\hbar$, $3\hbar/2$ and so on.

   For description of large numbers of particles, one needs to use statistical methods. It is shown in quantum mechanics that particles with a half-integer value of spin obey the Fermi–Dirac statistics and are known as "fermions," the particles with an integer value of spin obey Bose–Einstein statistics and are known as "bosons." The two families of particles have different roles in the world around us. A key distinction between the two families is that fermions obey the Pauli exclusion principle that was formulated by W. Pauli in 1925. According to this principle, there cannot be two identical fermions simultaneously in the limits of one quantum system, which means they cannot be present in the same place with the same energy, or in quantum language, they cannot have the same quantum numbers. In contrast, bosons have no such restriction, so they may "bunch" together even if in identical states.

6. In many cases, microparticles move with velocities that are close to the speed of light c. In this case, the rules of the theory of relativity should be considered. Such particles are called relativistic.

7. As a rule, SI units should be used exclusively in *books*, but in practice, in nuclear physics, SI units are never used to describe energy. In nuclear physics, the energy of particles E is mostly expressed in the extra-systemic unit "electronvolt" (eV). It is commonly used with the SI decimals prefixes – kiloelectronvolt (keV),

megaelectronvolt (MeV), and gigaelectronvolt (GeV). The particles with energies in fractions of eV are the subject of chemistry and thermodynamics. The particles with energies in tens and hundreds of eV are the subject of atomic physics. Nuclear physics begins from the energies of the order of 1 keV. The energies of nuclide sources, which determine radiation practically in all applications and radiation background as well, are in the range of MeV: fractions of MeV to several MeV.

8. Most of the known radiations are unstable particles that decay over time into stable ones. Almost all unstable particles have a very small span of life – small parts of seconds. Only a free neutron lives for an average 14.8 minutes (the period of half-life $T_{1/2} = 10.23$ minutes). Physicists know only several stable particles with very great, most likely infinite, life spans. They are as follows: electron, proton (i.e., the nucleus of the hydrogen atom), and heavier nucleus. The nucleus of the helium atom is the most well known among them. At last, photon must be pointed out here.

1.2 Constitution of Nucleus

The atomic nuclei consist of positively charged protons and electrically neutral neutrons. If the electric charge is neglected, the properties of the proton and neutron are so similar that they both are called by one name – *nucleon*.

The number of protons in a nucleus determines the electric charge and hence the number of electrons in an atom. The number of protons is numerically equal to the ordinal number of the element in the periodic system and is called the atomic number. Atomic number is denoted by the letter Z.

The number of neutrons is denoted by the letter N. The sum of the number of protons and neutrons $A = Z + N$ is called the mass number and approximately defines the nuclear mass.

There is a common designation of a nucleus in the form ${}^{A}_{Z}X$, where X is the name of the element. For example, ${}^{1}_{1}H$ is the nucleus of hydrogen, ${}^{235}_{92}U$ is the nucleus of uranium with the mass number 235 and so on. As the name uniquely identifies the atomic number, sign Z often falls in the designation, for example, ${}^{60}Co$, ${}^{137}Cs$, and so on. Sometimes the designation may be calcium-48 or Ca-48.

Nucleus with the same Z but different A are called isotopes, those with the same A but different Z are called isobars, while nuclei containing the same number of neutrons $N = A - Z$ are called isotones. A specific nucleus with given A and Z is called a nuclide. Nuclide is an official term defined by the standard [1].

In nature, many elements consist of a mixture of isotopes in definite concentrations. The majority of elements with odd atomic numbers have only one stable isotope. The elements with even atomic numbers have as a rule several stable isotopes. Further, we shall use the terms "nucleus" and "nuclide" for an individual nucleus and the terms "element" and "substance" for a natural mixture of isotopes.

In light nuclei, the number of protons is approximately equal to the number of neutrons, that is, $Z/A \sim 0.5$. For example, ${}^{4}_{2}He$, ${}^{12}_{6}C$ and so on up to ${}^{40}_{20}Ca$. Hydrogen is an exception, however, whose nucleus has only one proton, $Z/A = 1$. With the rise of Z, the number of neutrons overtakes the number of protons, experiencing minor fluctuations around the average from one element to the other and reaches the value $Z/A = 0.39$ for uranium. Such a relation between the number of protons and neutrons corresponds to the stable nucleus. If the relation of the number of protons and neutrons differs from the stable one, then the nuclei undergo radioactive decay.

In our world, the nuclei with all values of Z from 1 up to 118 are known (June 2014) but only nucleus with the atomic numbers up to $Z = 112$, 114, and 116 have the approved names. The region of stable nuclei ends up with bismuth ($Z = 83$). Then up to uranium ($Z = 92$), the radioactive elements follow. All elements with $Z > 92$ are artificially produced. They are called transuranic elements. There are no stable nuclei with $Z = 43$ (technetium) and $Z = 61$ (promethium). The stable nuclei, except $A = 5$ and $A = 8$, have mass numbers A from 1 up to 209. The production of more and more heavy transuranic elements continues. The heaviest nucleus produced up to now is Ununoctium ($^{294}_{118}$Uuo, $Z = 118$, $A = 294$). It is assumed that this nucleus is a member of the group of inert gases.

In Figure 1.1, the proton-neutron diagram of the known nuclei is presented. It is seen from the plot that for the light elements, the neutron/proton ratio is near 1:1. The more

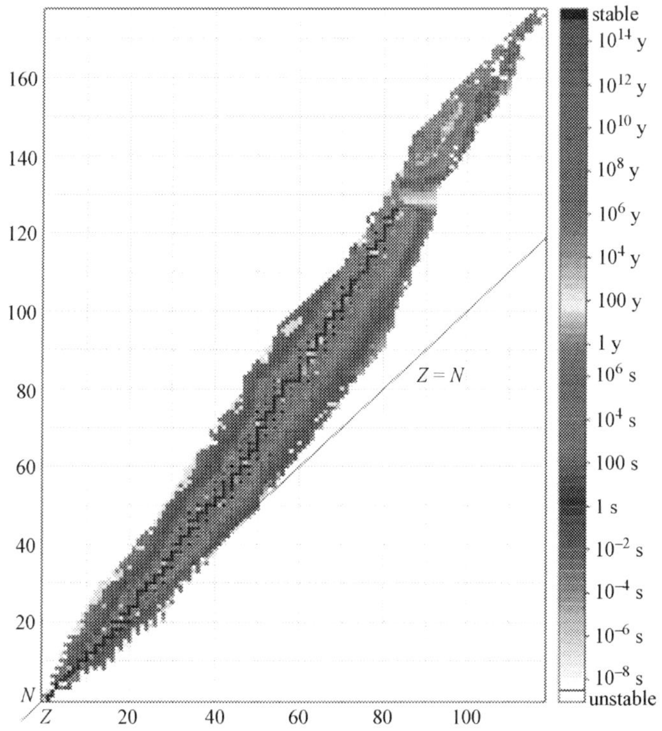

FIGURE 1.1 The proton–neutron diagram. Stable nuclides are shown by the black points in the middle of the grey region (colored region in the web version), that designates radioactive nuclides. Their half-lives decrease from the middle of the region to the edges (in the web version their stability is marked by color from red through green and blue to white with decreasing half-life) [2].

the atomic number, the more neutrons are demanded to compensate the repulsive force between protons. For the heavy elements such as uranium, this ratio is about 1:1.5.

1.3 Radioactive Decay and Radioactive Radiations

It has been said above that nuclei become unstable if the ratio of protons and neutrons differs from the definite range. All unstable nuclei, even in a ground state, undergo spontaneous transformation that got the name "radioactive decay." The analogous transformations undergo all excited nuclei.

There are several forms of radioactive transformations. Among the well-known and most probable are alpha decay, beta decay, and spontaneous fission. Gamma radiation is emitted by nuclei at transitions between energy levels without changing their composition. So the gamma transition is not, strictly speaking, a decay but, according to the tradition, sometimes the term "decay" is used in this case as well. All processes of decay are spontaneous but in the case of fission it is pointed out that this process is spontaneous in order to distinguish this relatively rare process from the main process in nuclear energetics – fission of nuclei under the action of neutrons.

The decays are more evident if they are shown in the energy diagram (Figure 1.2A–D). On the abscissa, the atomic number Z is meant, and the nuclear mass is represented on the ordinate (expressed in energy units, often arbitrary, not to scale). Since potential energy calibration, that is, zeroing is arbitrary, it is usually assumed that the energy of the ground state of the daughter nucleus (it is shown in the diagrams in Figure 1.2 by a thick line) is equal to zero. If the decay involves not only basic states of nuclei but also excited levels, that is also shown in the diagram. If the number of neutrons in the nucleus is greater

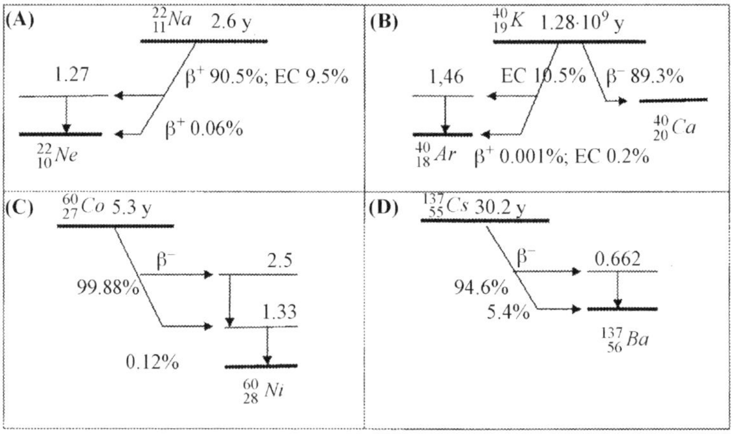

FIGURE 1.2 The energy diagrams of decays for several nuclides. (A) β^+ decay of ^{22}Na; (B) β^- and electron capture (EC) decays of ^{40}K; (C) β^- decay of ^{60}Co; and (D) β^- decay of ^{137}Cs. On the abscissa, the atomic number Z is meant, and the nuclear mass on the ordinate (expressed in energy units, not to scale). Arrows for β^- decay direct to the right, because nuclear charge in this case is increased. In the case of β^+ decay or the EC, the arrows direct to the left because nuclear charge decreases. Horizontal arrows indicate the energy level to which the transition occurs. Transitions between energy levels of the nuclei with the emission of gamma quanta are shown by vertical arrows.

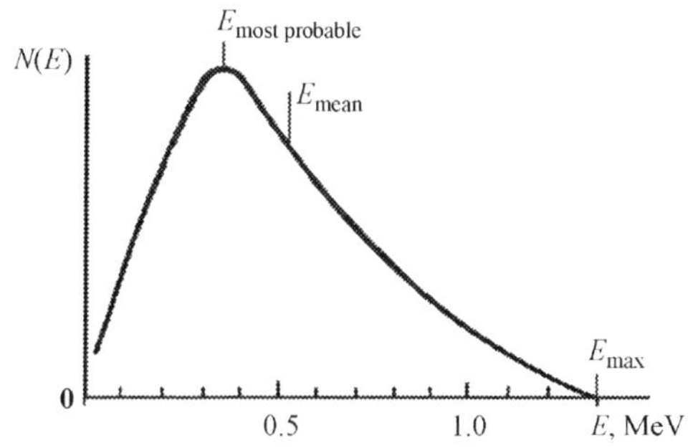

FIGURE 1.3 The spectrum of the beta decay of ^{40}K.

than the equilibrium value, then a β^- decay goes on, when a nucleus emits electron and antineutrino. If the number of neutrons is less than the equilibrium value, then a β^+ decay goes on. One option of β^+ decay is the electron capture (EC). In this case, a nucleus captures the electron of its own electron shell.

Often, in the process of beta decay, a nucleus transfers not to a ground but to an excited state and then beta decay is followed by gamma radiation. Transitions between energy levels of the nucleus with the emission of gamma quanta are shown by vertical arrows. Since nuclear levels are discrete and the gamma quanta carry away an overwhelming part of the nuclear excitation energy, the spectrum of gamma rays is discrete. Pure beta emitters whose beta emission is not accompanied by gamma quanta are quite a few, they are ^3H (tritium), ^{14}C, ^{32}P, and some others.

Beta decay produces electrons distributed in energy in the range from 0 to the maximum energy of decay of a given nuclide. Typical beta spectrum is shown in Figure 1.3. The maximum energies of beta decay may range from several tens of keV to several MeV. For example, $E_\beta^{max}(^3H) = 18.6$ keV, $E_\beta^{max}(^{14}C) = 156$ keV, $E_\beta^{max}(^{32}P) = 1.71$ MeV. Approximately, the same order of energy have gamma quanta that accompany the beta decay. However, in the case of gamma transitions, quanta with strictly defined energy are emitted. For example, ^{137}Cs: $E_\gamma = 662$ keV, ^{60}Co: $E_\gamma = 1.17–1.33$ MeV.

Fast electrons emitted in beta decay are called beta particles. If exactly the same fast electrons are produced in the other processes, they should not be called beta-particles but, simply, fast electrons.

The form of decay for heavy elements with $Z > 82$ is alpha particle emission and spontaneous fission. The typical energy of alpha particles emitted in the process of alpha decay is approximately 5 MeV.

1.4 The Radioactive Decay Law

The ability of a nucleus to decay is a property that does not depend on external conditions and on other nuclei. It is particularly important that the probability of decay does not depend on time, that is, a nucleus has no history. No matter how much time has passed since the formation of a radioactive nucleus, the decay probability per unit time is a constant,

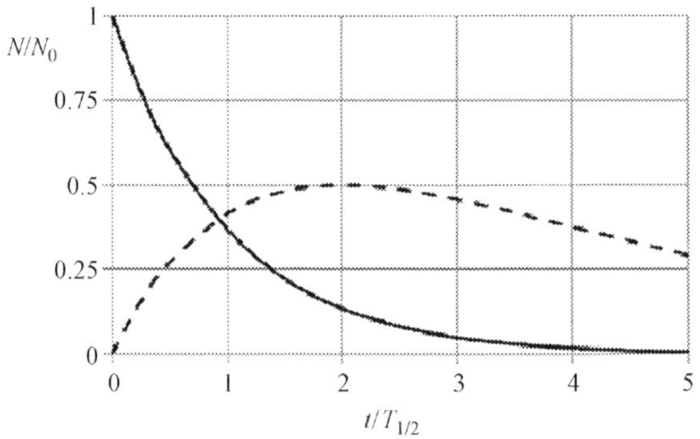

FIGURE 1.4 The law of radioactive decay. Solid line: the number of nucleus that do not decay till the moment *t*; dashed line: the number of radioactive daughter nuclei that are produced and then decay.

which is usually denoted as λ. From this condition, the well-known law of radioactive decay is derived

$$N(t) = N_0 \exp(-\lambda t). \quad (1.1)$$

Here N_0 is the primary number of nuclei, and $N(t)$ is the number of nucleus that do not decay till the moment t. This dependence is shown in Figure 1.4 (solid line).

The different nuclei from the number of similar nuclei live up to the decay for varying lengths of time. Average life span is equal to $\tau = 1/\lambda$. For historical reasons, in the science of radioactivity, the time during which the number of nuclei is reduced to half is used more frequent than the value of τ. This time is called the half-life $T_{1/2}$. The connection of $T_{1/2}$, τ, and λ is given by the following expression

$$T_{1/2} = \ln 2 / \lambda = 0,692 / \lambda = 0,692\tau. \quad (1.2)$$

The value dN/dt is the decay rate that is proportional to the number of existed nuclei N and the decay probability per unit time $-dN / dt = \lambda N \cdot$ The value λN is called activity of the definite radioactive chemical. Activity shows the number of decays in the unity of time. It is evident that the less the half-life, the greater the activity.

If the half-life is known, then the activity of the definite amount of a chemical can be lightly calculated. For example, for 1 gram of radium (^{226}Ra)

$$\lambda N = (0,692 / T_{1/2}) \cdot (N_A / A) = 3,7 \cdot 10^{10}\ \mathrm{s}^{-1},$$

where N_A is the Avogadro constant.

The activity of 1 gram of radium was measured by French scientists Pierre and Marie Curie at the very dawn of nuclear physics. This value in 1910 at the International Congress of Radiology, and electricity in Brussels was adopted as a unit of activity and was named "curie."

At present, the unit of activity in SI is a becquerel (Bq), named after A. Becquerel and equal to 1 decay per second.

One should note that according to the standard, activity is the number of decays per time unit rather than the number of emitted particles. The number of particles of a particular type is not necessarily equal to the number of decays. Therefore, such terms as alpha or beta activity, if they represent any number, must be understood as a tribute to history, but not taken as regular terms.

Often the decay of a nuclide (called the "parent") forms another ("daughter") nuclide that is also radioactive. Variation in the number of radioactive daughter nuclei that are produced, and then decay, is shown in Figure 1.4 (dashed curve). Provided that the half-life of the parent nuclides is much longer than all of the others', it can be shown that eventually an almost constant ratio between the numbers of nuclei of parent and that of all daughter substances can be reached. This ratio is equal to the ratio of their half-lives. The rate of decay of all substances becomes the same, and the number of nuclei decreases exponentially with the time of the parent substance, that is

$$\lambda_1 N_1 = \lambda_2 N_2 = \ldots = \lambda_n N_n. \tag{1.3}$$

In such situations, it is considered that the substances are in the radioactive balance. The relation (1.3) is called a secular relation.

1.5 The Radioactive Chains

There exist in nature three heavy radioactive nuclides with rather long half-lifes. So long that they had no time to fully decay since the formation of nuclides with which our planet, the Earth, is built, the so-called primordial nuclides. They are as follows: two isotopes of uranium ^{235}U–^{238}U and one isotope of thorium, ^{232}Th. Each of these nuclides is experiencing alpha decay; the daughter nuclide is also radioactive and it undergoes alpha or beta decay, transforming again into radioactive nuclide and so on until the last daughter (great-, great-…great-granddaughter) is appeared to be stable.

Such serial families of radioactive substances are called decay chains or radioactive rows. The decay of uranium-238 produces a chain that consists of 18 nuclides (uranium family, including starting material). The chain of thorium includes 12 nuclides (thorium family). All three of the above-mentioned decay chains end up by the different stable isotopes of lead.

To be more precise, we have to note that are several other long-lived nuclides known to us that keep staying on the Earth since the time of the elements' creation and have a $T_{1/2} > 10^9$ years. These are ^{40}K, ^{50}V, ^{87}Rb, ^{115}In, ^{123}Te, ^{138}La, ^{176}Lu, and ^{187}Re. All these nuclides are beta active. In the first decay of each of these nuclides, the stable substance is produced and they do not form radioactive chains. The most interesting among these is the isotope of potassium ^{40}K. This nuclide is presented in natural objects in the highest concentration. The decay scheme of this nuclide is shown in Figure 1.2.

Some characteristics of the radioactive chains of ^{238}U and ^{232}Th are shown in Figures 1.5 and 1. 6. In early studies of radioactivity, when real types of nuclides in chains were not yet determined, the chain members received special names. Since these names still appear

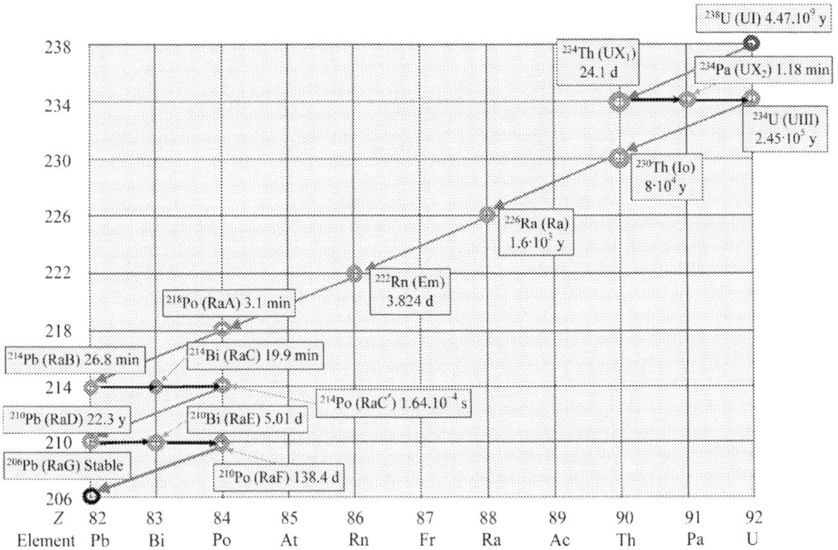

FIGURE 1.5 The uranium decay chain. The chain is shown in a simplified version; some transitions with small probability that form a "fork" at the end of chain are omitted. Gray arrows (red arrows in the web version): alpha decay; black arrows: beta decay. The first nuclide in the chain is marked by dark gray circle (red circle in the web version), the last one, stable nuclide – by black circle.

in the scientific and educational literature, not only the names of nuclides in their modern notation but also their old symbols are also presented on the schemes Figures 1.5 and 1.6.

It is seen from the schemes on Figures 1.5 and 1.6 that serial alpha and beta transformations go on in chains. In each alpha decay, the mass number decreases by four units and in

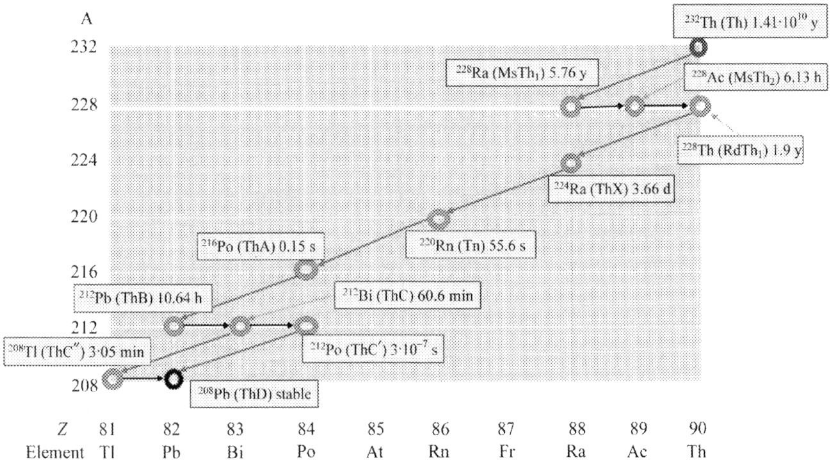

FIGURE 1.6 The thorium decay chain. In the end of the chain, one could see the so-called "fork." ^{212}Bi can undergoes alpha decay with the probability 36% and beta decay with the probability 64%. Gray arrows (red arrows in the web version): alpha decay; black arrows: beta decay. The first nuclide in the chain is marked by dark gray circle (red circle in the web version), the last one, stable nuclide – by black circle.

each beta decay, it does not vary. Therefore, the remainder of dividing the mass number A by 4 is the same for all nuclei of the same series. For a specific radioactive family, the value of A can be expressed by the formula $A = 4n + C$, where C is a permanent part of the family and n is an integer.

It is evident that only four radioactive rows can exist with $C = 0$, 1, 2, and 3. The thorium chain appears to correspond to the formula $A = 4n$, the uranium-238 chain corresponds to the formula $A = 4n + 2$, and the uranium-235 family corresponds to the formula $A = 4n + 3$. The fourth row with $C = 1$ contains nuclides, including the first nuclide of the row Np-237, that have small half-lives, decayed a long away, and now can be produced only artificially.

In the middle of the ^{238}U chain, as it is seen from Figure 1.5, there is the famous nuclide radium, ^{226}Ra. As the ratio of half-lives of radium and uranium is equal to $T_{1/2}(\mathrm{Ra})/T_{1/2}(\mathrm{U}) = 3.4 \cdot 10^{-7}$, the ratio of atom numbers must be the same. As mass numbers of these elements are close, 1 g of uranium contains $3.4 \cdot 10^{-7}$ g of radium, or 1 ton of uranium contains 0.34 g of radium. That's why the mining of radium is so difficult.

It is evident that the other nuclides of the chain, that has a much smaller half-life, except ^{234}U ($T_{1/2} = 2.45 \cdot 10^5$ y) and ^{230}Th ($T_{1/2} = 8 \cdot 10^4$ y), are present in the uranium ore in even smaller concentrations than radium. So their production would be even more laborious than radium, famous for its laborious mining. However, modern radiochemistry can separate such small amounts of radioactive substances.

1.6 X-rays

In Section 1.3 it has been told about gamma radiation was explained. In the processes discussed in this book, X-rays also play an important role. Gamma radiation and X-rays are the quanta of electromagnetic radiation that differ only according to their origin. Gamma quanta are generated inside the nuclei while X-rays are produced by electrons at the transitions between the deep inner atomic levels and as a result of electron deceleration in matter.

X-radiation was discovered by Wilhelm Röntgen, and in Russia and in Germany it is called "roentgen radiation." In English literature, it is called "X-ray" as Röntgen himself called it in order to signify an unknown type of radiation.

X-rays are produced in special X-ray generators. The simplest example of such generator is an X-ray tube. In the X-ray tube, electrons acquire energy in the electric field due to the potential difference between the cathode and the anode, and they bombard the anode (sometimes in the case of X-ray tubes, the anode is called "anticathode"). If bombarded electrons have enough energy, they knock out electrons from the deep electron shell.

As it is known, the energy states of the electrons in an atom are quantized and determined by a set of quantum numbers. Principal quantum number n determines the energy of the electron and its degree of distance from the nucleus, and can accept any integer values ($n = 1, 2, 3, \ldots$). Depending on the n, the electrons are arranged on the shells. The innermost shell corresponds to $n = 1$ and has a spectroscopic designation K-shell. Shell with $n = 2$ is called the L-shell and the rest follow in further alphabetic order.

The second quantum number l determines the orbital angular momentum of the electron. At the given n, the orbital quantum number l can accept integer values $l = 0, 1, 2, \ldots n - 1$. In the spherically symmetric Coulomb field (such a field only exists in the hydrogen atom), electron energy is determined only by the principal quantum number n and does not depend on l. In many-electron atoms, the field deviates from the spherically symmetric, and the energy of the electrons with different values of l also varies and depends not only on the principal quantum number n but also on the orbital quantum number l. Depending on l, each shell is subdivided into a certain number of subshells. In accordance with the quantum rules, the shell K has no subshells, shell L is split into three, shell M into five, and shell N into seven.

At the transitions between energy levels, monoenergetic quanta are emitted. The energy of this quanta are typical for atoms that emit them. The heavier the atom and the deeper the electron shell, the higher the energy of the characteristic quanta. Photon energy corresponding to transition of electrons from the L to the K shell of lead is equal to about 72 keV.

The transitions from various shells onto the K shell produce the K series, onto the L shell the L series, etc. The intensity ratio of the different lines is determined by the quantum rules, called the selection rules. As a rule, the brightest line in the K series is the K_α-line, and it is about 10 times brighter than the brightest line in the L series. Energy levels and transitions between them, leading to monoenergetic, characteristic X-rays are shown in Figure 1.7.

In a very short time, the vacancies at K shell, produced by bombarded particles, are filled by transitions from the outer shells, and energy equal to the difference between the energy levels is released. The release of energy can occur in two ways: either by X-ray

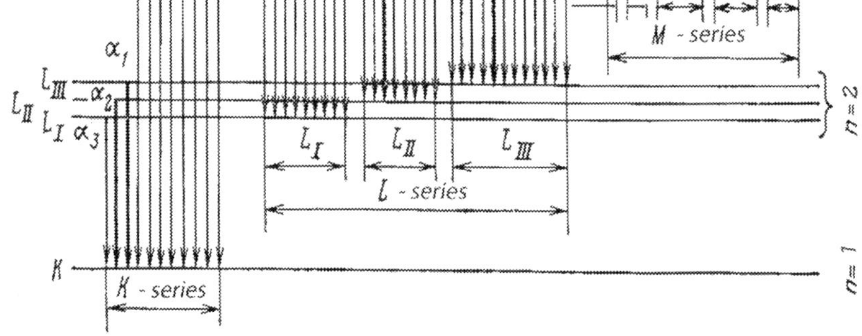

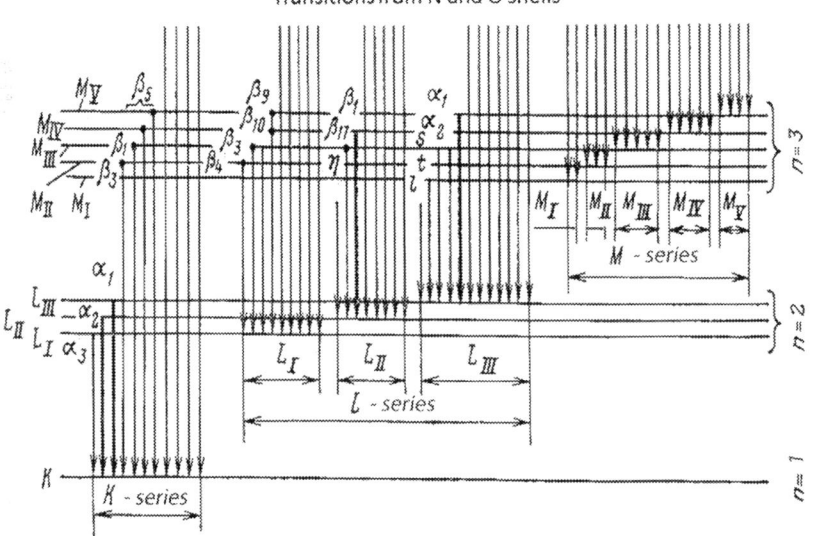

FIGURE 1.7 Energy levels of the inner atomic shells and the transitions that lead to emission of X-rays.

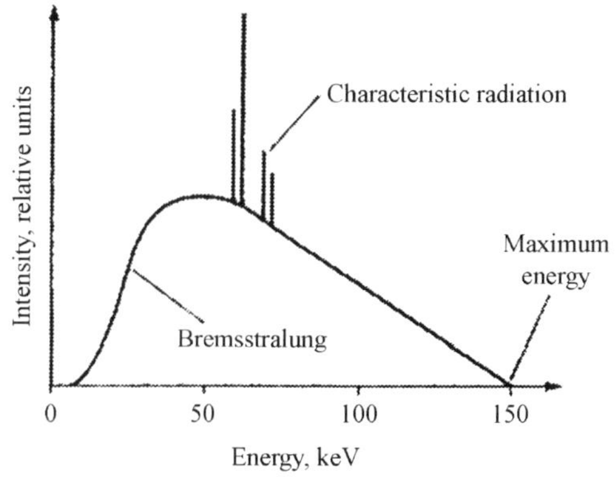

FIGURE 1.8 The typical spectrum of an X-ray tube with a tungsten anticathode.

emission, or when the transition energy is transferred to one of the electrons of the outer shell, which is emitted from the atom with a certain kinetic energy. Such transitions are called Auger transitions, and free electrons emitted from the atom, Auger electrons. In light elements, the Auger effect dominates. Beginning with $Z > 30$, the emission of X-rays becomes more probable.

Besides characteristic X-rays, the deceleration of electrons in matter induces the so-called bremsstrahlung radiation. The energy spectrum of this radiation is continuous; and the maximum energy in the spectrum is determined by the energy of decelerated electrons. The typical spectrum of an X-ray tube is shown in Figure 1.8.

On a scale of electromagnetic waves, the range of gamma quanta and the range of X-rays overlap. There exists soft gamma-radiation and hard X-rays. The boundary between these two types of waves is conditional. The typical value of X-ray energy for many radiobiological works is 250–300 keV.

1.7 Interaction of Ionizing Radiation With Matter

While analyzing the interaction of radiation with matter, it is convenient to divide all kinds of radiations into two groups – charged and neutral.

The group of charged radiations consists of all charged particles. This book is not concerned with great amounts of exotic shortlived particles that are produced in the conditions of the complex physical experiment, which only a limited number of physicists – specialists in high-energy physics – deals with. For the readers of this book, it is enough to be acquainted with some small set of charged particles that are really present in our life always and everywhere, regardless of our desire. Thus, here we consider the group of charged particles that includes fast electrons, protons, and the nucleus of heavier elements, mainly alpha particles. Besides, two types of mesons, pions (π^+, π^-, and π^0) and muons (μ^+, μ^-), play a certain role in the processes that are described in this book. They are produced in the atmosphere by primary cosmic radiation and are part of the cosmic rays bombarding all the living on Earth. Pions and muons are the most long-lived of all unstable families of mesons.

Passing through the matter, charged particles affect atomic electrons by their electric field and transfer them a portion of their energy. In so doing, they basically produce either ionization or excitation of atoms.

The basic characteristic peculiarity of charged particle passage through matter is the continuous gradual loss of energy in small portions. In the theory of particle passing through matter, there is even the accepted method of calculations called CSDA – continuous-slowing-down approximation. For example, an electron with energy 1 MeV in gaseous Ar at normal conditions has a range approximately equal to 4 meters and on this way, it produces ~38,000 ion pairs and ~11,000 excited atoms. It means that the distance between the nearest points of collision is slightly less than 0.1 mm. An alpha particle with energy 5.5 MeV in condensed matter, say, in water, has a range before it stops of ~45 μm, and on its way it experiences several hundred thousand collisions.

During the process of ionization, electrons can get the energy that exceeds the ionization energy. The remaining ionization electrons with excess kinetic energy are called delta electrons. Thus, the large energy of one primary particle is exchanged for a much lower energy of multiple electrons. About delta electrons, one can read more in detail in Section 4.1.1.

The group of neutral radiations consists of neutrons and gamma quanta. Unlike charged particles, neutral radiations experience rare collisions with atoms, in which they can lose either the whole or a substantial part of their energy. In these collisions, charged particles are produced that can affect the substance. Thus, neutral radiation converts into a charged one.

Ionization is appeared to be the most important and most evident process of interaction of nuclear radiations with matter. So they are called ionizing. Charged radiations are considered as directly ionizing and neutral ones as indirectly ionizing.

Charged particles passing through matter gradually lose their energy and eventually stop, that is, they have finite path in matter that is called range. Neutral ionizations collide with atoms randomly and the flow of radiation passing through the matter falls off exponentially. Thus, formally, the flow of neutral radiation reaches zero at infinity. So for neutral radiation, the concept of a range is replaced by the concept of an attenuation length.

All types of radiations finish their movement in matter till the practically full loss of energy within a time much shorter than the duration of all processes of energy conversion. Therefore, one can assume that the ionization of particles occurs almost instantly.

1.7.1 Specific Energy Loss, LET

One of the most important parameters that characterizes the interaction of charged particles with matter is the specific energy loss through ionization and excitation of atoms, energy loss on a unit of a particle path, dE/dx. The specific energy loss is sometimes called "stopping power"; this parameter contains information on how a particle loses its energy and how the matter stops it.

The simple analysis shows that the dependence of *dE/dx* on substance property is reduced to the relation

$$dE / dx \sim \rho, \tag{1.4}$$

where ρ is the substance density. It is convenient to divide both parts of relation (1.4) by density. In this case, one receives:

$$dE / \rho dx \sim \text{const.} \tag{1.5}$$

The term ρdx in the denominator shows that in nuclear physics it is more convenient to measure spatial values not in obvious linear units (i.e., in cm) but in mass units (i.e., in g/cm^2). The length expressed in mass units shows the amount of matter along the particle path. Specific losses expressed in mass units almost do not depend on the density of matter. They are strictly independent of the density for the same matter though they are slightly dependent on it, but close in value for different substances.

The dependence of specific losses on particle properties in a simplified way is expressed as follows:

$$dE / dx \sim \left(z^2 / v^2\right), \tag{1.6}$$

where z is the particle charge in the units of an elementary charge, v is the particle velocity. It is important that the specific losses do not depend on a particle mass. The dependence on mass will appear with transition from the dependence on velocity to the dependence on energy. In the relativistic case, the particle velocity is close to velocity of light ($v \sim c$) and practically does not change with the change of particle energy. Therefore, *dE/dx* is proportional to the square of the particle charge. In the nonrelativistic case, the numerator and denominator of equation (1.6) can be multiplied by the particle mass M, and in the denominator one obtains the particle energy

$$dE / dx \sim \left(Mz^2 / E\right). \tag{1.7}$$

It is seen that *dE/dx* is proportional to the mass and to the square of the particle charge and inversely proportional to the particle energy.

The relations (1.4–1.7) are approximate but allow to analyze the basic relations as they are. The real dependence of the specific losses on the particle energy is relatively complex, as it is shown in Figure 1.9.

In Figure 1.9, the energy of the particles is presented on the abscissa in units of the rest mass, that is, by the dimensionless factor E/Mc^2, and on the ordinate as the value $(1/\rho z^2)dE/dx$. In such coordinates, the curves of specific energy loss for different particles in different substances are approximately the same.

Protons, alpha particles, and the other nucleus at low energies can in a collision attach electrons from atoms of matter. As a result, they lose their charge, and hence the *dE/dx* decreases. The higher the probability to attach an electron, the less is the energy. Beginning

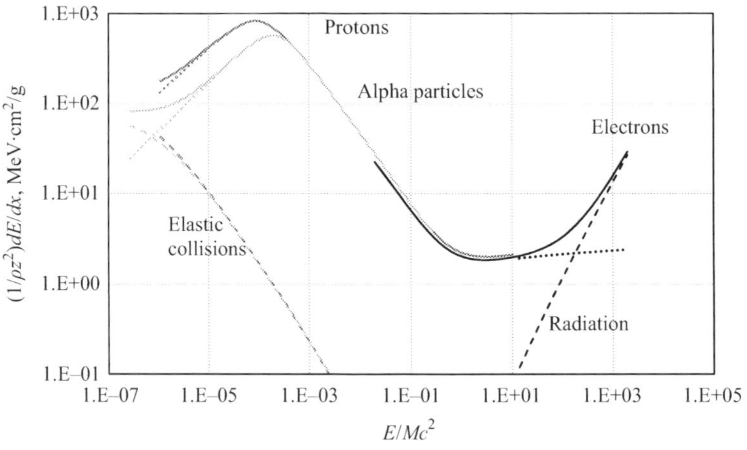

FIGURE 1.9 Specific energy losses for protons (dark gray [blue in the web version]), alpha particles (light gray [yellow in the web version]), and electrons (black [red in the web version]) in liquid water. Solid lines: total losses; point lines: ionization losses; dashed lines for protons and alphas: elastic losses; dashed line for electrons: radiation losses. The range of energy for protons: from 1 keV to 10 GeV; for alphas, from 1 keV to 1 GeV; for electrons, from 10 keV to 1 GeV. M_pc^2 = 938 MeV, $M_\alpha c^2$ = 3727 MeV, M_ec^2 = 0.511 MeV. The plot has been built by the author on the basis of data from the sites of National Institute of Standards [3].

with the lowest energy, specific losses increase and reach the maximum. The specific losses, expressed in corresponding coordinates, differ for protons and alpha-particles only in the range of electron attachment, as seen in Figure 1.9. The maximum value of LET for protons is achieved at E ~85 keV and is ~800 MeV·cm^2/g (80 keV/μm), and for alphas at 0.75 MeV and is approximately equal to 2200 MeV·cm^2/g (220 keV/μm).

Subsequently, with the rise of energy, dE/dx decreases and reaches the minimum. In Figure 1.9, it can be seen that $(1/\rho z^2)dE/dx$ does not differ for protons and alphas. The minimum of dE/dx corresponds to the particle velocity close to the velocity of light ($\beta = v/c \sim 0.96$) and to the particle energy $E \sim 2.5Mc^2$. After the minimum, the energy losses slowly, logarithmically, rise. In condensed matter, this rise is small and is several percents; in gases, it is higher – tens of percents. In estimated description of the passage of particles through matter, it is often assumed that the energy losses for all relativistic particles with the energy greater than the energy of a minimum are equal to minimum losses. For singly charged particles, the minimum specific losses in any matter is equal to ~2 MeV·cm^2/g (~0.2 keV/μm) for soft tissue and ~1.8 MeV·cm^2/g in air.

In Figure 1.9, elastic losses for protons and alphas and radiation losses for electron are also shown.

However, in analyzing the impact of particles on a substance, it is important to know how the substance gains energy rather than how the particle loses energy. One can find the answer to this question in the parameter that is called linear energy transfer (LET). This parameter is very similar to the specific losses and has the same signification. In the range of high energies, LET and dE/dx can differ rather significantly, but for low-energy particles, the difference is small or there is no difference at all. In general, LET is slightly smaller

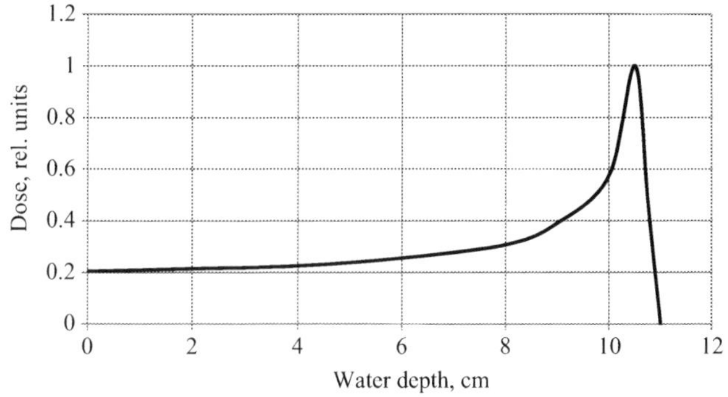

FIGURE 1.10 The Bragg curve for protons with energy 121 MeV in water. On the basis of [7].

than *dE/dx*. Within LET, the radiation loss and the part of the energy of high-energy delta electrons that can fly away from the particle track are not included.

In radiation physics, chemistry, and biology, it is conventional to express LET in kilo–electron volt per micrometer. LET of electrons and heavy particles differ very sharply. Typical values of LET are as follows: gamma radiation of ^{60}Co (i.e., electrons with energy near 1 MeV), 0.2–0.3 keV/μm; diagnostic X-rays with the energy 200–300 keV (i.e., electrons with the nearer energy), 2.5 keV/μm; fast neutrons (i.e., recoil protons), 60–70 keV/μm; alpha particles of radioactive nuclides, 90–100 keV/μm.

More detailed information about the track structure and LET values can be found in [4–6].

The passing of charged particles through matter has one important peculiarity. Let's have a look at the plot of the dependence of specific loss from a particle energy (Figure 1.9). In the figure, the particle begins its movement with some high energy and while moving in the matter it loses its energy and climbs up the curve of specific losses. Let's note that the fall of losses with decreased energy occurs in the far end of the particle range. As braking the energy loss per unit path of the particle grows. The main ionization effect the particle produces is at the end of its path. Such a feature in the behavior of particles was discovered shortly after the discovery of radioactivity by W. Bragg, senior. Dependence on the ionization (or energy loss, or absorbed energy) from the path of the particles in a substance, called the Bragg Curve, is shown in Figure 1.10 [7].

1.7.2 Multiple Scattering

In collisions, particles not only lose their energy but can change the direction of their motion. Very rarely, the particle undergoes such a collision in which the direction of the motion changes at once at a large angle. Such an event is called the single scattering. The single scattering exactly was the basis of discovery of the atomic nucleus. However, in each collision, the particle undergoes a very slight deviation, so weak that it is not possible to measure it experimentally. But a series of successive collisions can lead to the rejection of a particle on a completely measurable angle. This process is called multiple scattering. Root-mean-square angle of multiple scattering is the greater, the

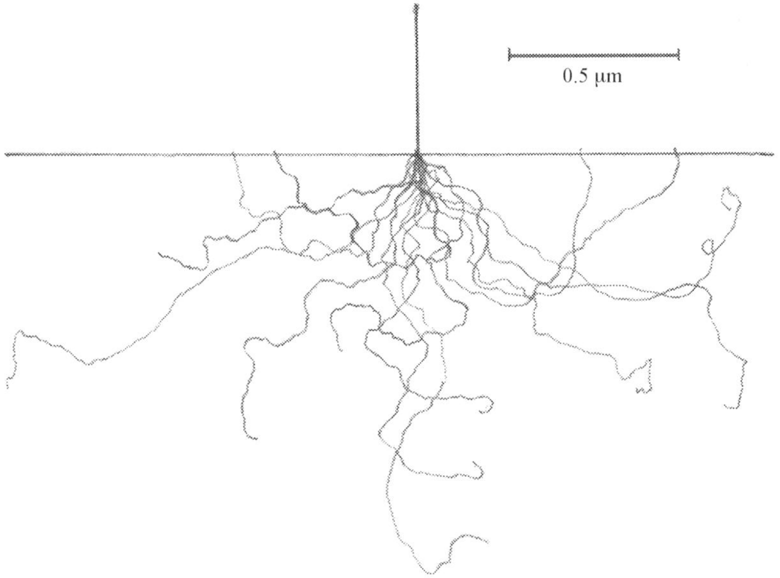

FIGURE 1.11 Monte Carlo simulation of the passing of electrons through matter that shows the role of the multiple scattering. Narrow beam of electrons with energy 20 keV incidents on the surface of an iron slab. Note that The CSDA range of 20-keV electrons in iron is equal to 1.77 μm [3].

lighter and the slower is the particle and the heavier is the substance. One example of the motion of light (electrons) and relatively slow (20 keV) particles in a dense substance (iron) is shown in Figure 1.11. Only several tracks from the great amount [8] are shown.

Because of multiple scattering, the path of a fast heavy particle in a relatively light material looks like a straight line; light tortuosity appears only at the end of the path. In the case of the motion of a light particle (electron) in a dense matter, the role of multiple scattering becomes very large (Figure 1.11). Quite quickly, the electron "forgets" the initial direction of motion. Its movement becomes chaotic and looks like a diffusion. The electron that enters such substance has a significant probability to change its direction of motion in such a way that it could fly out of the matter. In that case, electrons lose in matter only a part of their energy.

1.7.3 Particle Ranges

The path that a particle goes by in matter up to its complete stop, losing energy by ionization, is called range. Because of multiple scattering, the range appears to be quite a complex concept. Here only two parameters are considered: the full range along the curved path of the particle (CSDA range; see Section 1.7, the mean path length) and the distance in the direction of motion, which is called a practical range or sometimes an extrapolated range. Naturally the practical range is smaller than the full range.

The dependence of the range R, both full and especially practical, on particle energy E is not expressed in elementary functions. Usually, one uses empirical relations of the type

$$R = aE^b, \quad (1.8)$$

where a and b are constants. For electrons in water in the range of energy 500 eV–5 keV $b = 1.75$ and for $E > 5$ keV, $b = 2$. For alpha particles in air at energies that correspond to the energy of alpha particles of natural emitters ($E < 10$ MeV), $b = 1.5$.

1.7.4 Interaction of Gamma Quanta With Matter

The main modes of interaction of gamma rays with matter are the photo effect, Compton scattering, and electron positron pair production. In all these processes, charged particles – electrons and positrons – appear. Just these charged particles, which are produced while gamma quanta passes through matter, produce ionization and excitation.

Photo-effect is the absorption of gamma quanta by electrons, during which almost the entire energy of gamma quanta is transferred to the electron. Such process is impossible for the free electron because of violation of the law of momentum conservation. Therefore, the photo effect occurs on the atomic electrons and the probability of the process is higher the stronger the electrons are bound in atoms.

Gamma quanta can be scattered by the atomic electrons. If the gamma quantum energy exceeds the binding energy of the electron in an atom, then the gamma quantum can knock an electron out of an atom. In this process, gamma quanta lose their energy and change the direction of motion. This process is called the Compton effect. It was first demonstrated by the American physicist Arthur Compton who discovered it. For the discovery and interpretation of this effect, A. Compton was awarded the Nobel Prize for physics in 1927.

Gamma quanta with high energy ($E_\gamma > 2m_ec^2$) can produce electron–positron pairs, transferring its entire energy to the charged particles. This process is infeasible in vacuum and requires a nucleus or an electron in the neighborhood. Again as in the case of photo effect, it is impossible to observe the simultaneous fulfilment of the laws of momentum and energy conservation.

While interacting with matter, gamma quanta either transfer their entire energy to electrons or are scattered and eliminated from the beam. The attenuation of the gamma quanta beam is exponential:

$$n = n_0 \exp(-\mu x), \quad (1.9)$$

where μ is the total absorption coefficient that includes all three processes:

$$\mu = \mu_{\text{photo}} + \mu_{\text{Compton}} + \mu_{\text{pair}}. \quad (1.10)$$

The attenuation coefficients depend on the atomic number Z: $\mu_{\text{photo}} \sim Z^5$, $\mu_{\text{Compton}} \sim Z$, $\mu_{\text{pair}} \sim Z^2$.

In composite materials, the scattering and absorption of X-ray and gamma radiation relate to the effective atomic number (Z_{eff}) and the effective electron density.

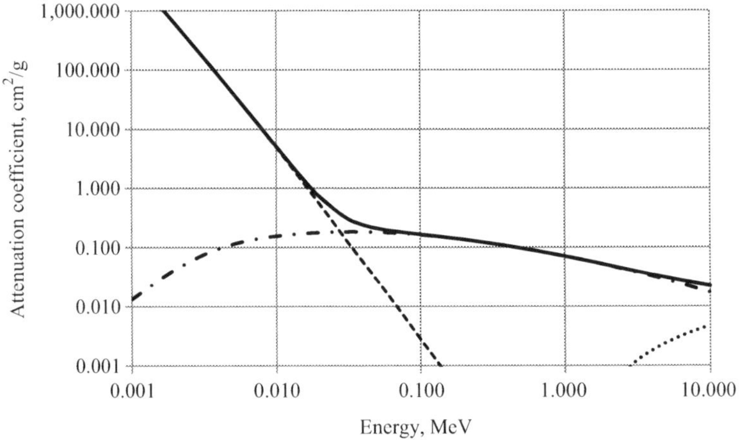

FIGURE 1.12 The attenuation coefficients of gamma quanta in water. Dashed curve: photo effect; dot-dashed curve: Compton effect; point curve: pair production; solid line: total interaction. The plot has been built by the author on the basis of data from the sites of National Institute of Standards [9].

The attenuation coefficients of gamma quantum interaction for all processes in water are shown in Figure 1.12.

The simple exponential attenuation of a gamma quantum beam in accordance with the formula (1.9) is valid only in the so-called geometry of the "narrow beam." In the Compton scattering, scattered gamma quanta can undergo a secondary scattering, and then a tertiary. As a result of a change of the direction of gamma quanta motion, a different number of gamma quanta than it is described by (1.9) can reach the x-coordinate. Therefore, for extended absorbers, one has to use effective absorption coefficients or build-up factors. [10].

Because of multiple Compton scattering, photons of different energies appear, when even monoenergetic radiation passes through matter. But the lower is the energy of gamma rays, the more likely their absorption is due to the photoelectric effect and elimination from the beam. Therefore, in the spectrum of gamma rays passing through matter, practically there are no low-energy photons. The spectrum of gamma rays passing through matter can be characterized by some average energy. If one considers the passing through air of the gamma quanta of nuclides of natural background radiation, that are usually present in the environment, the average photon energy in the air is equal to approximately 300 keV.

1.7.5 Interaction of Neutrons With Matter

Neutrons have no electrical charge, all their interactions are caused by the nuclear forces. Nuclear forces are the most powerful forces, but their range is very small and the neutron–nucleus collisions occur very rarely, much rarer than the collision of charged particles with atoms. Path of neutrons between collisions is in the range of centimeters or even more.

The physics of neutron–atom interaction principally differs from the physics of gamma-quanta interaction, but formally they are identical. Both gamma quanta and neutrons are penetrating radiations, and the attenuation of their flow is exponential. For both types of radiation, one can use the absorption and scattering coefficients.

A free neutron is an unstable particle. It undergoes beta decay with a half-life equal to 614 seconds. But all processes of neutron interactions end up in a time of much less than a second. So in further analysis, the instability of neutrons can be neglected.

As neutrons have no electric charge, they freely penetrate through atomic electron shells and are not pushed off by the electrical field of a nucleus. Therefore, neutrons are an excellent instrument with which one can study the nucleus, solid state, and biological structures and also produce new elements that are absent in the environment and can be very useful for medicine, industry, agriculture, and science.

The action of neutrons on matter occurs mainly through the charged particles produced in nuclear reactions which are induced by neutrons. In order to analyze possible ways of energy transfer from neutrons to charged particles, it is convenient to divide neutrons into two groups: slow and fast.

The most universal (runs on all nuclei, except ^{3}He and ^{4}He) exothermic nuclear reaction is the radiative capture (n,γ). Here in parentheses a particle that hits the nucleus and a particle that is emitted from the nucleus are written. The well-known example of the reaction (n,γ) is the reaction of radiative capture of neutron by the nucleus ^{113}Cd.

For the most part of the known nuclides, the cross sections of nuclear reaction are rather small. There are only several reactions with high values of cross sections. They are the reactions with the emission of alpha particle (n,α) with the nuclei ^{10}B and ^{6}Li and the reaction with the emission of proton (n,p) with the nucleus ^{3}He.

Heavy nuclei (^{235}U, ^{238}U, ^{232}Th, ^{239}Pu) can be split under the impact of neutrons. Some nucleus are split by the action of the neutrons of any energy, while others only by fast neutrons. As a result of fission, usually two fission fragments appear. The total kinetic energy of the fission fragments, e.g. for the fission of ^{238}U, is ~ 180 MeV. It is useful to recall that in one chemical reaction, the typical release of energy is on the order of electron volts.

There are a number of reactions in which the newly produced nucleus becomes radioactive with a significant half-life. Such reactions are called the ones with the induced activity. Depending on the initial nuclide, the new nucleus experiences β^-, β^+ radioactivity or electron capture (Section 1.3).

The probability of the most of nuclear reactions with neutron participation is maximal for the slow neutrons and decreases with the increasing of neutron energy.

One of the ground kind of interactions, which is essential for the influence of neutron irradiation on biological objects is the elastic scattering of neutrons by nuclei. This kind of interaction is important for fast neutrons. On collision with a nucleus, the neutron transfers a part of its energy, almost as the way it occurs in the collision of billiard balls. The lighter is the nucleus, the greater is the transmitted energy; therefore, the larger energy can get the nucleus of light elements constituting biological tissues – protons, in particular.

1.8 Elements of Dosimetry

1.8.1 Doses and Dose Rate

The impact of radiation on material objects is determined by the energy of the radiation transmitted to a substance and is measured by the value of the dose.

In the initial period after the discovery of X-rays and radioactivity, the main attention was focused on the effects of radiation on living organisms, primarily on the human being. Pretty quickly, in the first years, it was found that exposure to ionizing radiation leads to radiation damage. The kind of radiation damage first discovered was a radiation skin burn – erythema. To arrange a safe operation with radiations, it was required to establish a quantitative measure of the radiation field.

Historically, the so-called exposure dose, or simply exposure, was the first quantitative measure that was approved in 1928. It is based on the measurement of air ionization near the source of radiation, X-ray machine, or radioactive nuclide, because direct body measurements were not possible then. Exposure characterized possible effects of radiation on the soft tissues of the body pretty well, because the probability of absorption of photons depends on the atomic number (see Section 1.7.4), and the effective atomic numbers of air ($Z_{eff} = 7.64$) and soft tissue ($Z_{eff} = 7.42$) are pretty close.

The first unit of the exposure was a roentgen (symbol: R). In the Systéme International (SI), the unit of exposure is curies per kilogram (C/kg). It is important to note that the quantity exposure is only defined for X and gamma radiation and for the production of ions in air.

Subsequently, with increasing voltage on X-ray tubes and a corresponding rise of the radiation energy and greater use of radium, radiation can penetrate deeper into the body and affect the bone. It turned out that at the same magnitude of exposure, radiation effects are different in soft and in dense (bone) tissue due to the larger values of the effective atomic number of bone ($Z_{eff} = 13.8$). Also, the difference of density of soft tissue ($\rho \sim 1$ g/cm^3) and bone ($\rho = 1.85$ g/cm^3) plays a role.

The better way to assess the impact of radiation is the energy absorbed per unit mass of the substance. On the VII International Congress of Radiology, held in 1953 in Copenhagen, it was recommended to use the absorbed dose.

As the unit of absorbed dose, a rad was chosen (rad is the abbreviation of the words "radiation absorbed dose").

Closely related to absorbed dose is the quantity kerma, which is actually an acronym of "Kinetic Energy Released in MAtter." Kerma shows the sum of the initial kinetic energies of all the charged ionizing particles produced in the unit of mass. Kerma is a measure of energy liberated, rather than energy absorbed. The relation of absorbed dose and kerma is similar to the relation of the specific energy loss dE/dx, and linear energy transfer LET (see Section 1.7.1). The two will be equal under conditions of charged particle equilibrium, and assuming negligible losses by bremsstrahlung radiation. Kerma has the same units as absorbed dose.

For the soft tissue in the field of X-ray or gamma radiation, 1 rad approximately corresponds to the exposure equal to 1 R.

In 1960, the 11th General Conference of Weights and Measure (CGPM) approved the International System of Units (Système International d'Unités), usually known as "SI." Then the gradual introduction of international units in different countries and in different branches of science and economy begins.

The unit of absorbed dose and kerma in SI is Joules per kilogram (J/kg). This unit has a special name: gray (symbol: Gy), after Louis Harold Gray, a British physicist and one of the founders of radiobiology. This unit was adopted as part of SI by the 15th GCPM in 1975. The relation of the units is as follows: 1 Gy = 100 rad.

The rate of dose accumulation is called dose rate. It shows a dose of ionizing radiation delivered per unit time. For example, grays per second, per hour or per year.

Note that gray, as well as rad, evaluate the physical effects of the absorbed radiation, but do not take into account the characteristics of the biological effects occurring during irradiation. Therefore, it is necessary to introduce special units.

Biological effects of radiation on the body depend on the LET of radiation. The destructive effect of radiation is higher the more the LET. To account for this fact, the concept of equivalent dose was developed in the 1950s. The equivalent dose is calculated by multiplying the absorbed dose by a special factor w_R. This coefficient is called as either the coefficient of Relative Biological Effectiveness (RBE), the radiation weighting factor or the quality factor. On the difference between the definition of these coefficients, see Section 4.4.5. Here it is more correct to speak about the quality factor. The simplified table of the quality factors is given in Table 1.1 [11,12].

Much more detailed tables of the radiation weighting factors are developed. For example, to adequately calculate the equivalent dose in a mixed neutron field with the known neutron energy spectrum, one can use the table with 22 points with neutron energy from 0.025 eV up to 400 MeV, with a maximum value occurring in the range of 100 keV to 2 MeV. One can find them in the regulation documents [13] or, for example, in [14].

The recommended values of the radiation weighting factor have varied somewhat over the years, as evidence from biological experiments has been given and interpreted. The current values are per the recommendations of the International Commission on Radiological Protection (ICRP) [11].

However, even such specification of the concept of dose is not enough. Various human organs have different sensitivity to radiation (quality factors are usually defined on the basis of one biological effect, for example, such as reddening of the skin). Therefore, it is

Table 1.1 Radiation Weighting Factors w_R (on the Basis of [11,12])

Radiation	Photons	Electrons	Protons	α-Particles
Energy	All energies	All energies	>2 MeV	All energies
w_R	1	1	2	20

Table 1.2 The Tissue Weighting Factors w_T for Selected Organs and Tissues (on the Basis of [11,12])

Organ or Tissue	w_T	Sum of w_T Values
Bone-marrow (red), colon, lung, stomach, breast	0.12	0.60
Adrenals, gall bladder, heart, kidneys, muscle, pancreas, prostate (♂), small intestine, thymus, uterus/cervix (♀)	0.12	0.12
Gonads	0.08	0.08
Bladder, oesophagus, liver, thyroid	0.04	0.16
Bone surface, brain, salivary glands, skin	0.01	0.04
Total		1.00

necessary to introduce a special value, the effective dose, which considers the different radiosensitivity of different organs. Every organ of the body has its own tissue weighting factor.

To obtain an effective dose, the amount absorbed by the organ is corrected for the radiation type using the radiation weighting factors w_R, and then it is further corrected for the tissues or organs using the tissue weighting factor w_T. The sum of effective doses to all organs and tissues of the body represents the effective dose for the whole body.

The sum of the tissue weighting factors is equal to 1.0, so that if an entire body is radiated with uniformly penetrating external radiation, the effective dose for the entire body is equal to the equivalent dose for the entire body. If only part of the body is irradiated, then only those regions are used to calculate the effective dose.

The values of the weighting coefficients are defined empirically. The other name of these coefficients is coefficients of radiation risk. The values of the weighting coefficients of selected organs are shown in Table 1.2. The tissue weighting factors were revised in 1990 and 2007 because of new data.

Thus, equivalent dose is a radiation-weighted dose quantity, which takes into account the type of ionizing radiation producing the dose. Effective dose is a tissue-weighted dose quantity, which takes into account the different radiosensitivities of various organs and tissues.

The units of the measurement of both equivalent and effective dose are the same. Moreover, as the coefficients w_R and w_T are really dimensionless, so fundamentally, the units of the equivalent and effective doses are the same as absorbed dose. But these two types of doses show different effects, so they need different units. It was needed to reflect the biological effects of radiation as opposed to the physical aspects.

The old unit of measurement of the equivalent and effective doses was rem (rem is the abbreviation of the words Roentgen Equivalent in Man, or Mammal). In the former USSR (now Russian Federation), this unit before the year 1963 had the name "biological equivalent of roentgen–ber" and after 1963, "biological equivalent of rad-ber."

After introduction of SI, the rem was replaced by sievert (Sv), which was named after Rolf Maximilian Sievert, a Swedish medical physicist, one of the founders of radiobiology.

Originally, the unit rem was connected with roentgen. In 1977, the rem was redefined by the ICRP as 0.01 sievert or 0.01 J/kg, with the intention that the sievert would come to replace the rem. So older units of rem were smaller than the modern rem.

It is seen from Table 1.1 that for photons and electrons, numerical values of the absorbed dose in gray and the equivalent dose in sievert are equal. To get the equivalent dose for alpha particles and fast protons, one needs to multiply the absorbed dose by the radiation weighting factor.

After the introduction of SI exposure, and also exposure rate were not recommended for use, and therefore did not receive a special name. However, until the present time (July 2014) the units of exposure and exposure rate, based on the X-ray unit, are widely used. For example, on the official site of the Russian Federation national nuclear corporation Rosatom [15] dosimeter readings give the dose rate in **μR/hour and in μSv/hour.**

The Community Environmental Monitoring Program (CEMP) [16] gives information about the radiation situation in Nevada, United States, expressed in μR/hour.

The radiation situation in the United States and some other countries and continents on the site Radiation Network [17] is given in CPM (counts per minute). Conversion of these units depends on the type of dosimeter and the type of radiation [18].

In the use of units, the United States went its own way. It is well known that the United States uses so-called customary units that are virtually identical to the British imperial units, in its commercial activities and ordinary life, while science, medicine, government, and many sectors of industry use metric units. Nevertheless, the rem is still in common use, although regulatory and advisory bodies are encouraging transition to Sieverts [19].

Analysis of the risk of exposure to a large group of people – the whole population of a city, state, or all mankind – plays an important role for society. One can assess the risk for a whole group with the help of the collective dose. It is obtained by multiplying the mean effective dose with the total number of people who have been exposed to radiation. The SI unit of measurement of the collective dose is man-sieverts. The person-rem is sometimes used as the non-SI unit in some regulatory systems. The collective dose is the basis for the risk assessment of the effects of nuclear weapons testing, operation of nuclear facilities, and other sources of radioactive pollution in the environment.

1.8.2 The Connection of Radiometric and Dose Values

Connection of radiometric (source activity A) and dose (absorbed dose rate *P*) values is given by the relation

$$P = AeEk\mu / w4\pi r^2, \quad (1.11)$$

where k is quantum yield, w is the energy of formation of an ion pair, r is the distance from the source to the place of radiation absorption, μ is the absorption coefficient, and e is the elementary charge. If the source emits only one line of gamma rays, the absorbed dose is easily calculated from the above formula. But if it emits a few lines or even a lot, then the calculation becomes difficult because each radiation energy corresponds to its own

Table 1.3 Air Kerma Rate Constants and Exposure Rate Constants for Some Radionuclide Considering Photon Energy Above 20 keV (on the Basis of [20,21])

Nuclide	Air Kerma Rate Constant, aGy·m^2/(s·Bq)	Gamma-Constant, R·cm^2/(h·mCi)
^{22}Na	77.3	11.8
^{60}Co	84.98	12.96
^{137}Cs	21.77	3.32

absorption coefficient. In theory and practice of dosimetry, it is accepted to call the value $eEk\mu / w4\pi = \Gamma$ the Specific Gamma Ray Dose Constant of the nuclide, sometimes known as the Gamma Factor. Then the formula (1.11) is significantly simplified

$$P = A\Gamma / r^2. \tag{1.12}$$

The values of the gamma constants for different nuclides are tabulated.

Before the introduction of SI, the gamma constant characterized the exposure of gamma radiation point isotropic radionuclide source, determined for the following standard conditions: source activity = 1 mCi, and the distance from the source to the detector $r = 1$ cm, that is, the dimension of gamma constant, was R·cm^2/(h·mCi). Now for the same purposes, the constant kerma is used. In this case, the activity is measured in becquerel, dose in gray, and the distance in meters. The dimension of kerma rate constant is Gy·m^2/(s·Bq). Considering that typical doses in gray are expressed in a small number, and the activity in becquerels, in very large doses, it is convenient to use kerma constant in dimension aGy·m^2/(s·Bq), where the prefix "a" (atto) is a factor 10^{-18}.

The values of the kerma rate constants and gamma constants for some nuclides are presented in Table 1.3 [20,21].

It is useful to point out the other characteristics of particle flow. Particle flux is the number of particles incident on all areas of an absorber in a unit of time, and the possible dimension is s^{-1}. Particle fluence is the number of particles incident on a unit of area for all time of measurement, and the possible dimension is m^{-2} or cm^{-2}. Particle fluence rate is the number of particles incident on a unit of area in a unit of time, and the usual dimension is cm^{-2}·s^{-1}. Sometimes one needs to use the quantity particle radiance that is the fluence rate of particles propagating in a specified direction within a stated solid angle, and the possible dimension is cm^{-2}·s^{-1}·sr^{-1}.

1.8.3 Microdosimetry and Nanodosimetry

When analyzing the effects of ionizing radiation on biological structure, the subject of analysis has microscopic sizes, it is cell or cell nucleus, or even a single molecule. In this case, the quantum nature of the energy loss of the charged particles and the associated statistical nature of the distribution of energy plays an important role. Analysis of fluctuations of energy is the subject of a special branch of science – microdosimetry.

Microdosimetry is a branch of radiological physics that provides quantitative characterization of the non-uniformity of energy deposition in uniformly irradiated matter [22].

Microdosimetry typically operates with the specific energy parameter Z – the ratio of the energy transmitted to the small volume of the substance to the weight of this volume. The probabilistic nature of absorption leads to fluctuations in value Z, which is the larger, the smaller the volume, the lesser the radiation dose, and the more the LET of ionizing particles. This fact can be illustrated with the following example [23]. At a dose of gamma radiation 10^{-2} Gy, the deviation of Z from the average Z_{av} in the volume of a living cell is approximately 10%. In the case of neutron radiation with the same dose (ionization is produced by recoil protons, and LET is considerably larger), the fluctuations are such that in 9 cells from 10, $Z = 0$, and in every 10th cell, Z can exceed Z_{av} 10 times. In the volume occupied by the chromosome (at the same dose of neutron radiation) $Z = 0$ in 999 chromosomes from 1000, and in one of them Z may exceed Z_{av} in 1000 times. Specific energy fluctuations are significant for volumes with linear dimensions of about 1 μm.

In recent years, one can observe a further advance into the depths of matter, as nanodosimetry is coming to move microdosimetry. Nanodosimetry is the next logical extension from microdosimetry, providing information on radiation track structure at a DNA or nanometer level [24].

The opportunity to provide measurements of radiation events in nanometer-sized volumes appeared sometime after 1995 but it was the dream of microdosimetry for several decades [25,26].

1.9 Radiation Detection

Currently, in the arsenal of physicists, there is a wide range of instruments for the detection, research, and dosimetry of ionizing radiation [27,28]. However, here we confine ourselves to the two most common and the most important methods of detection in the context of this book – ionization and scintillation methods.

A charged particle moving through the matter ionizes atoms. This matter, called usually as working media, can be a gas, liquid, or solid. If the working media is situated between electrodes with a voltage applied between them, then electrons and ions (or holes in semiconductors) produced by charged particles form in the external circuit electric current, which is registered. This is the basis of the ionization method of particle detection.

In some materials, electrons and ions (holes) recombine efficiently with the emission of photons of light. Furthermore, charged particles produce not only ionization but excitation of atoms as well. Quanta of light may occur during transitions of excited atoms in the ground state. Registration of photons using photosensitive receivers is the basis of the scintillation method of particle detection.

The value of signal from the ionization detector per 1 keV of particle energy is ~50 electrons from the gas and ~500 in semiconductors, that is, 0.5–5 μV in a standard circuit. Sensitivity threshold of modern devices for particle detection can be taken as about 0.1 V. Thus, the signal from the detector should be amplified. There are two principal possibilities to amplify the signal – either directly in the detector, when one can use some physical processes to amplify the signal, or external amplification, which can be done with the help of electronic devices.

One of the main ways of internal amplification of the signal is the creation of electron avalanches in the gas of a detector. Under certain conditions, these avalanches can become so powerful that the detector develops self-sustained gas discharge.

First, with the help of self-sustained gas discharge, a single alpha particle was detected in 1908 by H. Geiger and E. Rutherford. They used the discharge electrode system tip-plane. In 1928, H. Geiger together with his PhD student W. Müller significantly improved the device, replacing the tip by a thin metal wire. Since that modification, the Geiger-Müller counter has been an indispensable tool for measuring radioactive radiation.

Electron avalanches in a Geiger counter are so powerful that the counter "forgets" about what particle with what energy produces its operation. In principle, a Geiger counter outputs a signal with the standard amplitude even if a charged particle produces in a volume just one electron. That is, the Geiger counter is just a counter, it detects particles completely independent of their type and energy. Therefore, a Geiger counter cannot be the recording device of a dosimeter. Widespread domestic dosimeters that are based on Geiger counters are actually radiometers and determine the dose very conditionally. They allow to make relative measurements but are not suitable to compare readings between different radiation detectors.

Most recently, on the market quite simple dosimeters based on semiconductor detectors have appeared.

Scintillation detectors that use the registration of photons with the help of some kind of photosensitive sensors, particularly photo-electron multipliers, possess very good detection properties. Now one can observe the replacement of vacuum photomultipliers by multi-pixel avalanche semiconductor photon sensors (multi-pixel photon counter [MPPC]), which are significantly easier to manufacture and cheaper. It can be expected that after a while, scintillation dosimeters will be as compact and cheap as dosimeters with gas detectors.

1.10 Natural Radiation Background

All life on Earth is immersed in the ocean of ionizing radiation. Life came into existence in the irradiation conditions, probably more intense than in the current era, developing under the conditions of exposure and reaching the modern state under irradiation, and it continues to exist, still immersed in this ocean. These radiations fall on the surface of the Earth from space, and come from the radioactive substances in the earth's crust, and from the buildings around us, the food, water, air, and even in our body.

All radiation sources can be divided in several groups:

1. The sources of natural origin (85%) and resulting from human activities including nuclear medicine (15% on average); in developed countries and especially in the USA, the situation is quite different (see Table 1.4);

Table 1.4 Approximate Relation of Contributions of Various Sources Into Total Annual Dose (on the Basis of [29–31])

	USA		Russia	
Source	**Dose, mSv/year**	**%**	**Dose, mSv/year**	**%**
Natural sources				
Radon (internal, only lungs and bronchi)	2.3	37	1.7	43.1
Cosmic (space, external)	0.31	5	0.3	7.6
Terrestrial (soil, external)	0.19	3	0.35	8.9
Internal (food and water, mainly ^{40}K and ^{14}C)	0.3	5	0.38	9.6
Total natural sources	3.1	50	2.73	69.2
Artificial sources				
Medicine[a]	3.0	48	1.2	30.5
Others[b]	0.1	2	0.01	0.3
Total external irradiation	0.6		0.66	30.8
Total	6.2	100	3.94	100

[a]Medicine includes computed tomography (24%), nuclear medicine (12%), interventional fluoroscopy (7%), and conventional radiography/fluoroscopy (5%).

[b]Others include coal-fired power plants, nuclear power plants, fertilizers, consumer products, atmospheric nuclear testing, nuclear accidents, nuclear fuel cycle.

2. The sources of the earth origin (92%) and cosmic (8%);
3. The sources of external (35%) and internal (65%) irradiation.

The approximate ratio of the contributions of different sources to the total average annual dose is shown in Table 1.4 [29–31].

For selected persons and groups of population, different sources of background radiation can vary significantly depending on the place of residence and type of activity. For example, the annual dose due to terrestrial sources for Louisiana residents is 0.15 mSv, while for Colorado residents it is 1.4 mSv. In Europe, the average natural background exposure by country ranges from under 2 mSv annually in the United Kingdom to more than 7 mSv annually in Finland. The contribution of medical applications of nuclear techniques is continuously growing for the population of the United States. Now (July 2014) this medical component reaches an average of 3 mSv per year across the U.S. population (global average = 0.6 mSv/year).

Typical average normal background radiation in the open countryside, expressed in units of absorbed dose rate, is ~1 mGy/year = 0.1 rad/year. Often radiation dosimeters show radiation background in μR per hour. For rough estimates, one can take 1 mGy/year to 10 μR/hour. This background includes only external irradiation and does not include internal exposure due to radon and radioactive nuclides that enter the body with food.

The radiation background on the territory of Russia can be found on the website Rosatom [15] in the section "Radiation Map." There one can find the readings of dosimeters installed at various locations within nuclear power plants and in plants for fabrication and processing of nuclear fuel.

Let's analyze some of the sources of background radiation.

1.10.1 Cosmic Radiation

One of the important sources of ionizing radiation are cosmic rays penetrating everything on Earth invariably all time of its existence.

Physicists know a large number of elementary particles, but the vast majority of them are short-lived, whose lifetime is a small fraction of a second. Only stable particles reach the boundary of the earth's atmosphere from space. They are called the primary component of cosmic radiation. The majority of them are protons (90% of the total particle number); alpha particles (~7%); and heavier nuclei, mainly C, N, and O (~1%, but heavier nuclei up to uranium are also still detected). Moreover, primary cosmic rays contain a quantity of electrons and positrons (~1%) and gamma-rays (about 0.01% when the energy >100 MeV). Primary particles come mainly from the outer space, the average "age" of cosmic rays is estimated at 20–100 million years. Some of the primary particles come from the sun.

The total particle fluence rate of cosmic particles on the boundary of the atmosphere during the solar activity minimum is $\sim 1\ cm^{-2}\ s^{-1}$ and decreases two to three times approaching the maximum.

Primary gamma and X quanta are totally absorbed by the earth's atmosphere. Their detection and investigation is the subject of X-ray and gamma-astronomy and are possible only on the spacecraft and high-altitude balloons.

High-energy particles of the charged component of primary cosmic rays interacting with atoms in the atmosphere create secondary cosmic radiation. Near the earth's surface, the primary radiation is approximately 1% of the total flow of the particles. The total particle fluence rate of cosmic particles at sea level is $\sim 1\ cm^{-2}min^{-1}$.

Neutrons, born in nuclear interactions in the atmosphere, are captured by nitrogen nuclei in the reaction $^{14}N(n,p)^{14}C$ and by oxygen nuclei in the reaction $^{16}O(n,\alpha)^{14}C$, creating radioactive nuclide ^{14}C in the atmosphere. This nuclide has a half-life of 5730 years. Nuclides ^{14}C accumulate in the atmosphere to some equilibrium concentration and participate in carbon circulation.

Because of cosmic radiation, dose rates increase with altitude: background sea level, 0.03 mSv/hour (~0.26 mSv/year); 2000 m above the mean sea level, 0.1; 4000 m, 0.2; 12000 m, 5; and 20,000 m, 13 mSv/hour. In highland cities of Bogota, Lhasa, and Quito, the level of cosmic radiation is about five times higher than that at sea level. In particular, the cosmic background for residents of Denver, Colorado, is about 0.5 mSv/year, which is almost two times higher than that for residents of Boston, on the shore of the ocean.

Cosmic rays produce an increased background in aircraft cabin. Thus, the average effective dose during a transatlantic flight is about 5 μSv/hour. Over a 10-hour flight, the accumulated dose is about 0.05 mSv. Especially, large doses accumulate in crew members

Table 1.5 Content of Uranium, Thorium, and Potassium-40 in Some Natural Objects

	U	Th	^{40}K
Granite, ppm	3–4	14	4
Upper layers of soil, ppm	3–4	8-10	Strongly depends on fertilizer
Ocean water, g/L	$(2–3)\cdot 10^{-6}$	$1\cdot 10^{-8}$	$4.5\cdot 10^{-5}$
Organism of an adult man with mass 70 kg	$7\cdot 10^{-4}$ g (~10 Bq)	$7\cdot 10^{-5}$ g (<1 Bq)	$3\cdot 10^{-3}$ g (4–5 kBq)

of airliners – about 1.6 mSv/year. For comparison, a typical chest X-ray, 0.02 mSv; mammography, 3 mSv; and a single session of whole body CT scan, 45 mSv.

1.10.2 Terrestrial Radiation

In the time since the formation of the Earth, all the short-lived nuclides are decayed and only long-lived nuclides with half-lives of the order of billions of years left in the rocks. They include two isotopes of uranium ^{235}U and ^{238}U, one isotope of thorium ^{232}Th, and radioactive isotope of potassium ^{40}K (about the others, see Section 1.5).

Natural radioactive substances uranium, thorium, and potassium-40 are contained in many objects that surround humans, particularly in seawater, soil, stones, etc., as well as in building materials and determine a natural radioactive background. The concentration of radioactive nuclides in some objects is given in Table 1.5.

Increased concentration of thorium in granite provides additional background. So according to the website of Rosatom [15] radiation background on the granite embankment of the Neva in St. Petersburg is 45 μR/hour.

It is interesting to point out that relatively high amount of ^{40}K presents in banana. So one banana with mass 150 g has an activity of about 20 Bq.

The presence of uranium and thorium in the natural objects determines by gamma-radiation not of nuclides U and Th themselves but by the lines emitted by a member of the corresponding chain that are the most convenient for measuring.

In the family of thorium, the nuclide ^{208}Tl (ThC"), standing at the end of the chain, has in its gamma spectrum the line 2.62 MeV. Energy of gamma radiation of all radioactive nuclides, either natural or appearing as a result of human activity, is lower than this value. Therefore, the line 2.62 MeV can easily be separated on the level of radiation of the other substances.

The presence of uranium (^{238}U) is usually detected by the lines 1.76 or 2.2 MeV that ^{214}Bi (RaC) emits.

1.10.3 Some Special Radionuclides: ^{222}Rn, ^{3}T, ^{14}C, ^{85}Kr

1.10.3.1 Radon, ^{222}Rn

Radon occurs in the alpha decay of radium; it is radioactive, and undergoes alpha decay with a half-life $T_{1/2}$ = 3.824 days. Radon is a member of the 8th column of the periodic table, which are inert gases. If the decay of radium occurs near the surface of the material, the atom of radon can go out into the air, take part in the movement of air as one of its constituents, and accumulate, especially in indoor air. A typical value of radon activity in indoor air is approximately 100 Bq/m^3, but can grow tenfold in unventilated areas and in some other cases. Inhalation of radon produces about half of the natural human dose and is the largest component of the source of one nature. For more information about radon, see Section 4.6.

In the thorium family, there is also the isotope of radon, which is called thoron (^{220}Tn). The half-life of thoron is equal 55.6 seconds, it cannot essentially move out of the point of its appearance and does not practically accumulate.

1.10.3.2 Tritium, ^{3}T

$T_{1/2}$ = 12.32 years, the energy of the decay is 18.59 keV, the average energy of electrons is 6.5 keV. Tritium is generated in the upper layers of the atmosphere as a result of the interaction of cosmic rays mainly with nucleus N and O, during tests of nuclear and mostly thermonuclear weapons and in nuclear power plants. In the atmosphere, tritium quickly oxidizes and turns out in the molecule of superheavy water HTO. According to current estimates, the equilibrium activity of cosmogenic tritium in the environment is $(1.11–1.30)\cdot 10^9$ GBq (3.0–3.5 kg). About 90% of natural tritium is contained in the hydrosphere. The explosion of a hydrogen bomb with a TNT equivalent of 1 MT leads to the release of $(2.6–7.4)\cdot 10^8$ GBq of tritium (0.7–2 kg of tritium). Modern nuclear power plants annually emit several dozen kilograms of tritium. Tritium is present in the human body, where it enters with the water, food, inhaled air, and through the skin. The average human body contains about $5\cdot 10^{-12}$ g of tritium, which makes a very small contribution to the total dose.

1.10.3.3 Carbon-14, ^{14}C

Carbon is the main element of all biological processes. Natural carbon consists predominantly of the stable isotope ^{12}C. The isotopes ^{10}C and ^{11}C have half-lives 19 seconds and 20.5 minutes, respectively, and ^{14}C of 5730 years. In the decay, ^{14}C emits only beta particles with the maximum energy of 154 keV. The isotope ^{14}C is widely used in radiocarbon dating.

Carbon-14 is generated in the atmosphere on interaction of cosmic rays with atmospheric nitrogen and is rapidly oxidized to a state of $^{14}CO_2$. As a result, an equilibrium concentration is established. In preindustrial period, the concentration ratio $^{14}C/^{12}C$ was equal to $1.2\cdot 10^{-12}$. This concentration ratio characterizes organisms that consume carbon dioxide during perspiration. When the organism dies, the carbon exchange stops, new portions of ^{14}C does not arrive, and it just decays. Subsequently, the concentration ratio falls. By measuring the residual concentration of ^{14}C, one can calculate when the organism died. This is the basis of radiocarbon dating.

Atmospheric nuclear weapon tests in the 1950s and 1960s of the last century led to a rapid increase in the concentration of ^{14}C. In 1963, the concentration of atmospheric ^{14}C exceeded pre-atom era by about twofold. After the ban of nuclear tests in the atmosphere, ^{14}C content began to decline and now it exceeds the level before testing to about 10%.

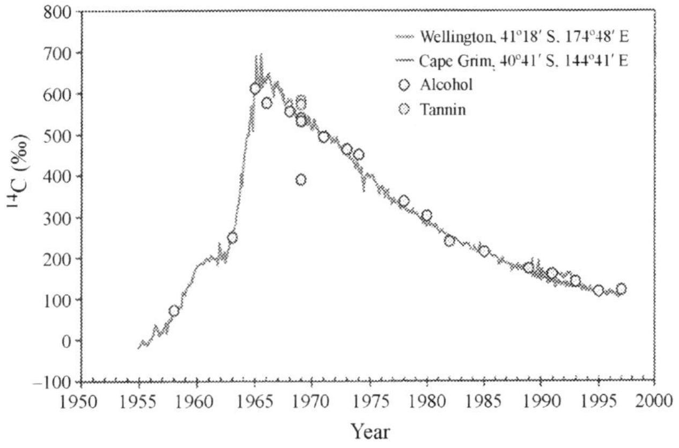

FIGURE 1.13 Radiocarbon concentration in vintage red wines. From [31] with permission of IAEA.

Impact of nuclear explosions on the content of ^{14}C shown in Figure 1.13 [32]. The levels of ^{14}C were obtained from measurements of atmospheric CO_2, tree rings, many organic objects, and ice layers.

The observed change in the content of ^{14}C in the atmosphere found very interesting criminological applications. Earlier, radiocarbon dating was used for time periods whose duration was comparable with a half-life, that is, times from a few thousand to about 50 thousand years. However, the discovery of concentration jumps associated with testing of nuclear weapons (bomb pulse) permits to date recent biological materials.

Hence, measuring the concentration of ^{14}C in wine, one can determine if the wine is old, that is, produced before 1950s, as the manufacturer insists, or is it falsification.

Many artificial nutritional supplements are prepared from petroleum-based materials. These materials do not contain radioactive carbon. A natural product contains it in concentrations typical for the time of production. Consequently, by analyzing the concentration of ^{14}C, one can determine the proportion of artificial additives in food product.

In the fight against drug trafficking, this method was used to determine the age of the drug to understand if there is a fresh production or somewhere there is a store of drugs extracted earlier. It is enough to have milligrams of substance for analysis.

^{14}C dating has allowed to estimate how much new growth of nerve cells occurs in human adults. People who were alive during the testing of nuclear bombs above ground took up more of ^{14}C and incorporated it into the DNA of their cells. Using accelerator mass spectroscopy to count carbon-14 atoms in the DNA from hippocampal nerve cells from the autopsied brains of 55 people who lived during that time, the team of scientists found that about 35% of the nerve cells in the hippocampus had been renewed over the course of adulthood [28].

Currently $^{14}C/^{12}C$ ratio decreases exponentially with a half-life of about 15 years. After a few years, seasonal variations exceed deviations, and the dating of fresh biological materials in such a way would be impossible.

1.10.3.4 Krypton-85, ^{85}Kr

This is the isotope of inert gas that has a relatively long half-life, $T_{1/2}$ = 10.8 years. ^{85}Kr is practically a pure beta emitter, the energy of decay is equal to 0.687 MeV, and the average electron energy is 0.251 MeV.

The main sources of ^{85}Kr in the atmosphere are nuclear weapon tests, work of nuclear power plants, accidents, on nuclear enterprises. During the period of nuclear tests in the atmosphere (1945–1962), approximately $190 \cdot 10^{15}$ Bq ~ 5 MCi has been injected into atmosphere. The accident at the Three Mile Island nuclear power plant in 1979 injected approximately $1.9 \cdot 10^{15}$ Bq and the Chernobyl accident 10 times more. The maximum concentration of ^{85}Kr in the atmosphere has been achieved in the 1970s (about 0.4 Bq/m^3), and has since decreased.

A large nuclear power plant in operation annually produces about $1 \cdot 10^{16}$ Bq ^{85}Kr; basically, it is stored in the rods of the fuel elements, but a certain part still goes out. Additional intake of ^{85}Kr in the atmosphere occurs during the processing of nuclear fuel.

References

[1] IUPAC (1997). "Nuclide." In: McNaught AD, Wilkinson A, editors. Compendium of Chemical Terminology. Blackwell Scientific Publications. http://old.iupac.org/publications/compendium/index.html.

[2] Plot of atomic isotopes. http://commons.wikimedia.org/wiki/File:Isotopes_and_half-life_1.PNG.

[3] National Institute of Standards. Physical Measurement Laboratory. Stopping power and range tables: for electrons http://physics.nist.gov/PhysRefData/Star/Text/ESTAR.html, for protons – http://physics.nist.gov/PhysRefData/Star/Text/PSTAR.html, for alphas – http://physics.nist.gov/PhysRefData/Star/Text/ASTAR.html.

[4] Mozumder A. Charged Particle Tracks and their Structure. Adv. Rad. Chem. 1969;1:1–102.

[5] Mozumder A, Hatano Y, editors. Charged Particle and Photon Interactions with Matter: Chemical, Physiochemical and Biological Consequences with Applications. Boca Raton, FL: CRC Press; 2003. 860 p.

[6] Hatano Y, Katsumura Y, Mozumder A, editors. Charged Particle and Photon Interactions with Matter – Recent Advances, Applications, and Interfaces. Boca Raton, FL: CRC Press; 2011. 1045 p.

[7] Zhang X, et al. Parametrization of multiple Bragg curves for scanning proton beams using simulteneous fitting of multiple curves. Phys. Med. Biol. 2011;56:7725–35.

[8] Site of the Tomsk Polytechical Institute. Lectures for students – in Russian - http://portal.tpu.ru/SHARED/c/CHERDANTSEV/gh/Tab1/lkc.ppt.

[9] National Institute of Standards. Physical Measurement Laboratory. XCOM: Photon Cross Sections Database http://www.nist.gov/pml/data/xcom/index.cfm.

[10] Dosimetry & Shielding - http://www.nucleonica.net/wiki/index.php/Help:Dosimetry_%26_Shielding.

[11] ICRP. The 2007 Recommendations of the International Commission on Radiological Protection. Ann. ICRP. 2007 103 **37**(2-4).

[12] Wrixon AD. New ICRP recommendations. Review. J. Radiol. Protect. 2008;28:161–8. http://iopscience.iop.org/0952-4746/28/2/R02/pdf/0952-4746_28_2_R02.pdf.

[13] Current U.S. regulations.(Title 10, Code of Federal Regulations, Part 20, i.e., 10CFR20).

[14] Stabin MG. Radiation Protection and Dosimetry. An Introduction to Health Physics. Springer, 2007. Ch. 5, Quantities and Units in Radiation Protection, p. 72, Table 5.2.

[15] Rosatom - http://www.rosatom.ru/en/ - in English. Click to the icon "Radiation map" and you get to the site "Radiation situation on the enterprises of Rosatom – http://www.russianatom.ru - in Russian.

[16] Community Environmental Monitoring Program (CEMP) - http://cemp.dri.edu/cemp/.

[17] Radiation Network - http://radiationnetwork.com/.

[18] Converting from CPM to mR/hr - http://www.blackcatsystems.com/GM/converting_CPM_mRhr.html.

[19] Nuclear Regulatory Commission. "NRC Regulations: §34.3 Definitions." Retrieved 2007-03-14 - http://www.nrc.gov/reading-rm/doc-collections/cfr/part034/part034-0003.html.

[20] Unger LM, Trubey DK. Specific Gamma-Ray Dose Constants for Nuclides Important to Dosimetry and Radiological Assesment. Oak Ridge National Laboratory, ORNL/RSIC-45/R1, 1982.

[21] Wasserman H, Groenewald W. Air kerma rate constants for radionuclides. Eur. J. Nucl. Med. 1988;14:569–71.

[22] Rossi HH. The Role of Microdosimetry in Radiobiology. Radiat. Environ. Biophys. 1979;17:29–40.

[23] Ivanov VI, Lystsov VN, Gubin AT. Reference Guide on Microdosimetry. Moscow, 1986 – in Russian (Иванов В. И. Лысцов В. Н., Губин А. Т. Справочное руководство по микродозиметрии, М., 1986).

[24] Rabus H, Nettelbeck H. Nanodosimetry: Bridging the gap to radiation biophysics. Rad. Measurements 2011;46(12):1522–8.

[25] Grosswendt B. Introduction to Nanodosimetry - http://www.eurados.org/~/media/Files/Eurados/events/Winter_schools/2_Grosswendt_introduction%20to%20nanodosimetry.pdf.

[26] Grosswendt B. Nanodosimetry, from radiation physics to radiation biology. Radiat. Prot. Dosimetry. 2005;115:1–9.

[27] Bolozdynya AI, Obodovskiy IM. Detectors of Ionizing Particles and Radiations. Principles and Applications. Dolgopruny, Publ. House "Intellect", 2012, 204 p. – in Russian (А.И. Болоздыня, И.М. Ободовский. Детекторы ионизирующих излучений. Принципы и применения. Долгопрудный: Изд. Дом Интеллект, 2012, 204 с.).

[28] Spalding K, et al. Dynamics of Hippocampal Neurogenesis in Adult Humans. Cell 2013;153(6):1219–27.

[29] Sources of irradiation of Russia population (in Russian) http://www.ibrae.ru/russian/chernobyl-3d/man/1.htm.

[30] U.S. Environmental Protection Agency. Radiation Protection. http://www.epa.gov/radiation/understand/perspective.html.

[31] United States Nuclear Regulatory Commission. Fact Sheet on Biological Effects of Radiation. http://www.nrc.gov/reading-rm/doc-collections/fact-sheets/bio-effects-radiation.html.

[32] Tuniz C, Zoppi U, Hotchkis MAC. Bomb pulse radiocarbon dating. Analytical application of nuclear technique. Vienna, Austria: IAEA; 2004. p. 145–58. http://www-pub.iaea.org/MTCD/Publications/PDF/Pub1181_web.pdf.

[33] Knoll G. Radiation Detection and Measurement. 4th ed. John Wiley & Sons; 2010. 830 p.

2 Basics of Biology

The objective of this chapter is to highlight the concepts and phenomena that are essential for a better understanding of material presented later in this book. This proposed description of cell structure and certain cell processes does not aim to replace a systematic molecular biology course. Many fine details of intracellular processes are omitted; those interested can consult the list of references [1,2].

2.1 Cell Structure

The main targets of radiation and chemical effects are the body cells. The cell is the basic unit of structure and function of all living organisms. Apparently, the English scientist Robert Hooke was the first one to observe cells and give them their names in 1665, using a microscope invented in the late sixteenth century.

Understanding the emergence of the first cell is one of the most difficult and interesting problems of natural science. It is believed that cells formed about 3.5–4 billion years ago in the primordial soup, of which the seas in those days consisted. There are reasons to believe that the ancient Earth atmosphere contained sufficient concentrations of simple building blocks such as carbon dioxide, methane, ammonia, and hydrogen. From these components with minimal additives, all living things eventually got "built." Note that at the time, there was no oxygen in the atmosphere.

In modern times, four major classes of intracellular small molecules (amino acids, nucleotides, sugars, and fatty acids) can be obtained from the primary molecules, under the influence of ultraviolet irradiation and/or electric discharge, in a rather simple experiment (known as the Miller-Yuri experiment) [3]. One can presume that in the atmosphere of ancient Earth, this could happen too. These small molecules can associate, forming large polymeric chains as a result of spontaneous aggregation. Thus, proteins and nucleic acids are formed. The resulting organic substances started to concentrate in certain places, probably in drying foam on the shores or in shallow waters.

At some point, a particularly remarkable molecule randomly formed from existing components. R. Dawkins called it a replicator [4]. It had an unusual feature – the ability to create copies of itself. In principle, such a molecule should have formed only once. But to be sure, for safety one should assume that such an event could occur repeatedly. Later, a competition between the descendants of these first replicators started, and there were winners and losers.

Apparently, all modern cells are descendants of a single primitive cell line that has been able to "develop" an effective mechanism of protein synthesis. Although the picture of life's origin on Earth is not yet clear, it is conceivable that the first living cells appeared in a natural way, when a group of already quite complex organic molecules got surrounded by a shell – a membrane, that is.

It is rather difficult to imagine a spontaneous emergence of such an intricate system as a living cell. Therefore, alternative views are widespread. Here are two of them: panspermism, live cells having been brought from other worlds; and creationism, the divine creation of life as a result of a Supreme Being's intelligent design.

Obviously, both of these theories do not solve the problem. Panspermism transfers the emergence of life on Earth somewhere into the infinity of the universe. However, the question remains of how life originated there. Creationism leaves the decision with a Supreme power but does not explain how it was done, declaring the problem fundamentally incognizable. Science, of course, cannot be satisfied with this. Science requires knowledge.

2.1.1 Prokaryotes and Eukaryotes

The first cells were quite simple; they did not have a nucleus and are therefore called prokaryotic or simply prokaryotes. In these cells, in a single volume the RNA molecules (see below) stored genetic information and catalyzed the synthesis of necessary proteins. Such cells have survived to the present day – they are bacteria.

Approximately 1.5 billion years ago, there was a transition from prokaryotes to much larger and much more intricate cells that contained a nucleus in their structure. Now, the storage of genetic information was passed to the more reliable DNA molecules (see below) and became separated in a special compartment – the nucleus – from the rest of the vital processes of the cell, where this information could be damaged. Such nuclear cells were named eukaryotes. These are the cells of higher plants and animals.

Eukaryotic cells are larger than prokaryotic cells. Their diameter is about 10 times longer; therefore, the volume is approximately 1000 times greater. Typical dimensions of prokaryotic cells are 0.5–5 μm and those of eukaryotic 10–50 μm. Certain eukaryotic cells may have a very large size. For example, one ovule of a bird is its whole egg yolk. Neurons of large mammals can reach tens of centimeters in length.

A eukaryotic and a prokaryotic cell is shown in Figure 2.1 [5]. On the picture, only the most important parts of the cell are indicated. Further, we need to pay special attention to the following components: the nucleus, membrane, cytoplasm, and ribosomes.

A eukaryotic cell has a very complex structure. Free-living eukaryotic cells show an extraordinary diversity of forms. They may have photoreceptors, sensitive bristles, or various appendices enabling purposeful movement, etc.

Monocellular organisms, bacteria, and animalcular eukaryotes comprise more than half of the total biomass of the Earth.

However, evolution has not chosen the path of increasing a single cell's size and complexity but one of cell division and the division of responsibilities between different types of cells. It proved more beneficial to build multicellular organisms whose cells, as a result of differentiation, become specialized to perform specific functions and to form corresponding organs.

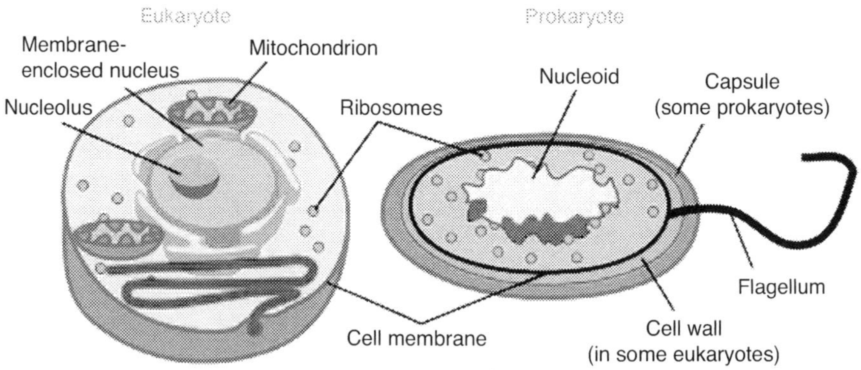

FIGURE 2.1 The cells of eukaryotes (left) and prokaryotes (right). Only most important parts of the cells are shown. From [5].

2.1.2 Nucleus, Chromosome, DNA, and Gene

The central role in a cell (often also by location, but mainly ideologically) is occupied by the nucleus. It is the nucleus that contains chromosomes – the media for storing information on the structure, chemical composition, behavior, and development program of the cell. Inside the nucleus the basic genetic processes occur. A significant part of the cell is filled with cytoplasm. The nucleus is separated from the cytoplasm by a special nuclear membrane, thus representing a deeply hidden and well-protected box. The typical size of a nucleus is 5 μm.

A chromosome is a composite of a deoxyribonucleic acid (DNA) molecule and special proteins (histones). The chromosome is so named because it is a body (soma) that can easily be stained by special colorants, which facilitates its observation under a microscope.

A DNA molecule, the main carrier of genetic information, is a linear polymer built of monomers called nucleotides. A nucleotide consists of a nitrogenous base, 5-carbon sugar, and one or more phosphate groups. In DNA and RNA, monophosphates are used. Non-organic phosphate is represented by a stable negative ion formed from phosphoric acid, H_3PO_4. The phosphate gives a negative charge to the nucleotide.

Bases in the DNA structure can be of only four types: two bases are purines: adenine (A) and guanine (G); two bases are pyrimidines: thymine (T) and cytosine (C). These compounds are called base because in an acid medium they behave as a base, that is, they can attach an ion (H^+).

In fine chemical experiments using chromatographic measurement (1949–1951), Erwin Chargaff with associates showed that samples of DNA from different biological sources keep to the same principle: the amount of adenine equals the amount of thymine, and the amount of guanine is the same as that of cytosine, that is, A = T, G = C. Additionally, the DNA contains equal amounts of purines and pyrimidines A + G = T + C. However, the ratio of (A + T)/(G + C) can vary significantly with different DNAs.

A 5-carbon sugar in DNA is deoxyribose. Sugar + base form nucleosides, which are called, respectively, adenosine, guanosine, cytidine, and thymidine. Nucleoside + monophosphate form nucleotides.

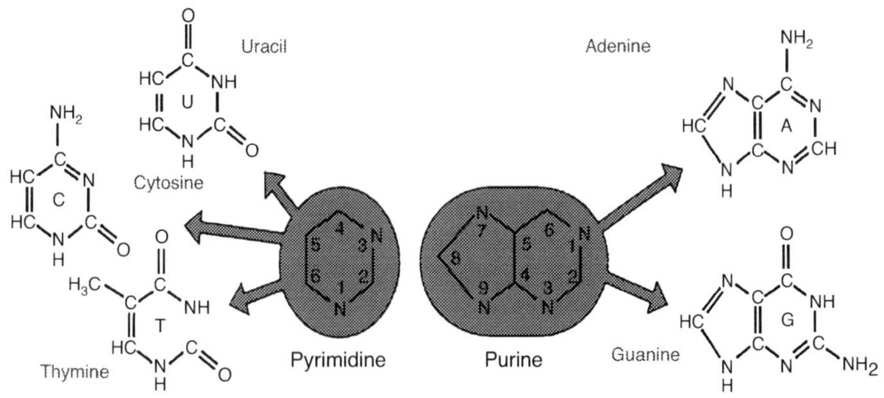

FIGURE 2.2 Structural formulas of the bases. The numbering of atoms in base cycles is shown.

Structural formulas of the bases are shown in Figure 2.2. The figure shows the numbering of the atoms in the base cycles. The connection of nucleotides into the polymer chain of DNA is shown in Figure 2.3. Here one could find the numbering of atoms in sugar–deoxyribose cycles. To distinguish between atomic numbering in sugar and that in the bases, sugar atoms are numbered on the outside of a cycle and a stroke (') is added to the figure (1', 2', 3', 4', and 5'), as it is shown in Figure 2.3.

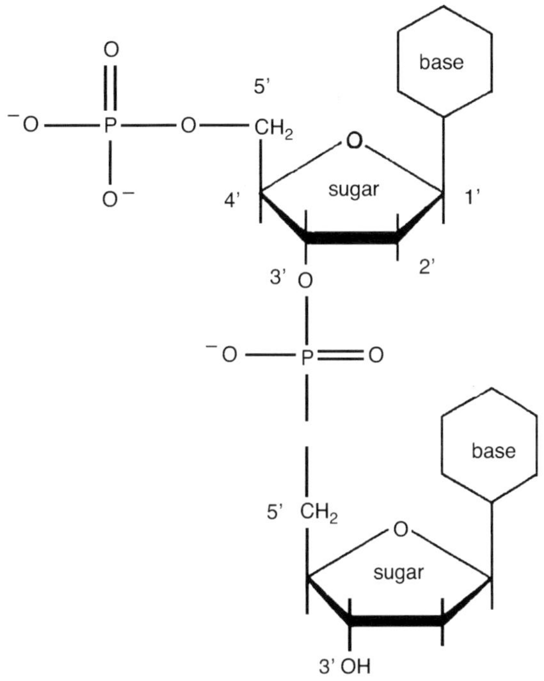

FIGURE 2.3 The connection of nucleotides into the polymer chain of DNA. The numbering of atoms in sugar cycles is shown.

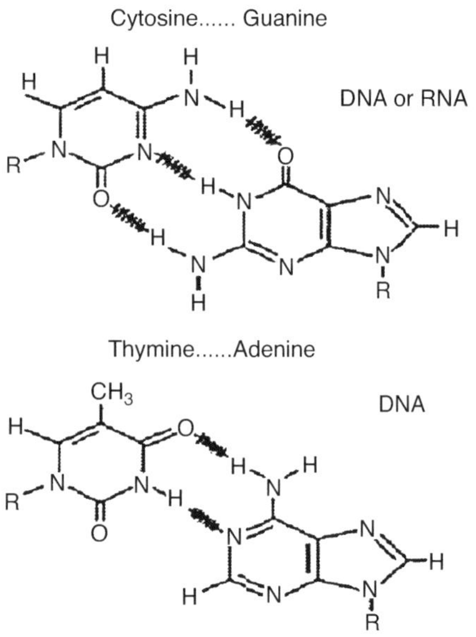

FIGURE 2.4 Hydrogen bonds between bases in DNA chains.

The DNA molecule consists of two polymer chains. The chains' framework is made of a sugar-phosphate chain; the chains are joined by mutually corresponding pairs of nitrogenous bases. As a result, in a single DNA molecule, purine is bound with pyrimidine: A is bound with T and vice versa, G is bound with C and the other way around. Bases are connected by hydrogen bonds. Between adenine and thymine (AT), there are two hydrogen bonds, and between guanine and cytosine (G–C), three, as shown in Figure 2.4. Therefore, the bond A–T can be broken somewhat easily than the bond G–C. Bond lengths between the bases are about 0.28–0.29 nm. The hydrogen bond in a guanine–cytosine pair is ~0.3 eV per bond (~7 kcal/mol). In normal conditions, the double helix is very stable. Conjugated bases are connected so firmly that in order to separate the two DNA strands (such a process in biology is called denaturation or melting), the temperature in a test tube needs to be very high. The relation between the degree of linking in a DNA helix and the temperature can be described by a smooth curve, but the range of temperature at which the separation of the chains occurs is small. The temperature at which the chain is half-split is approximately 85 °C. But special protein enzymes easily cope with this task in genetic processes, separating the DNA chain and bonding them back together.

DNA is curled into a right-handed helix. The diameter of double helix is 2 nm; the distance between adjacent base pairs is 0.34 nm. The double helix makes a full turn every 10 pairs. Each base pair has the same width and height.

The DNA length depends on the organism. The DNA of simple viruses contains only a few thousand components, that of bacteria, a few million, and those of higher organisms, in billions.

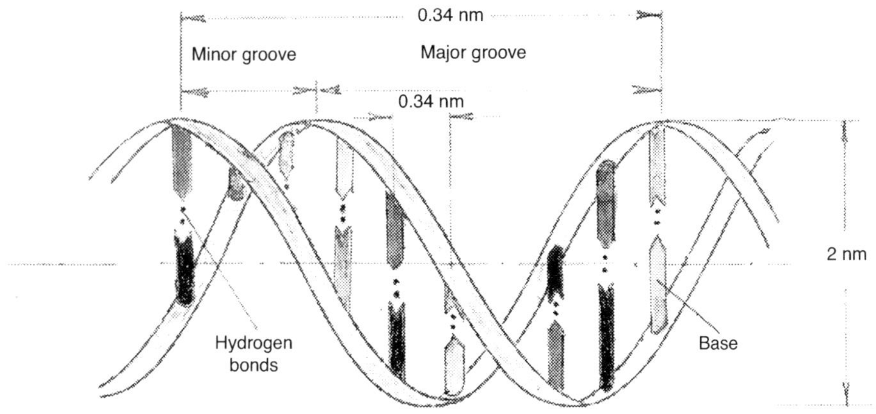

FIGURE 2.5 Schematic representation of DNA.

A schematic representation of DNA is shown in Figure 2.5. It can be seen as a spiral staircase whose steps are base pairs.

The structure of the DNA molecule and the way it functions were discovered in 1953 by J. Watson and F. Crick, who used X-ray diffraction patterns of M. Wilkins and R. Franklin and E. Chargaff's rule. In 1962, this achievement was recognized by awarding the Nobel Prize to Watson, Crick, and Wilkins. Rosalind Franklin had died by that time, and according to the rules the Nobel Prize does not get awarded posthumously.

A remarkable property of DNA is that some of the nucleotides comprising the polymer chains are capable of selectively binding to certain other nucleotides. As a result, extended sequences form, in which only certain pairs will be associated with another. This pairing property is called complementarity.

The DNA molecule's double-helix model is amazingly beautiful. Having given a huge boost to the development of biology, it has become an icon not only in biology textbooks but also in art [6]. However, it is useful to know that it is not the helicity that is important for understanding many of the processes but the double-stranded structure and the complementarity of the bases' bonds. These properties allow the DNA to replicate genes in a stable and reliable manner. Therefore, many of the processes associated with DNA can be shown on a flat picture of the molecule, as is done, for example, in the figure of the replication fork (Figure 2.6).

In total, human cells have 23 pairs of chromosomes, that is, 46 chromosomes. Further, 22 pairs of chromosomes are numbered according to the order of 1–22, and also X and Y chromosomes. In each cell of the body, or as biologists say, in somatic cells, there is a full set of chromosomes, that is, they are diploid. In female cells, a set of two X chromosomes is present; male cells have one X and one Y chromosome. Gametal cells (ovum and sperm cells) contain a single set, that is, only 23 chromosomes. An ovum has only X chromosomes, and a sperm either X or Y. Different organisms have different numbers of chromosomes and there is no real connection with the complexity of the organism. Thus, a human has 46 chromosomes, a dog 78, a drosophila 8, and a garden snail 54.

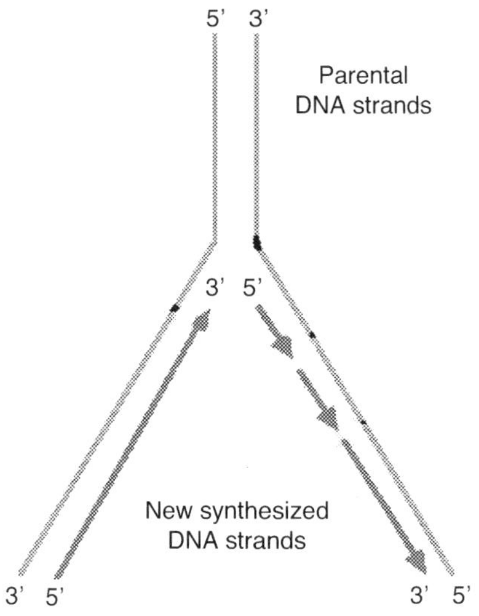

FIGURE 2.6 Replication fork.

The molecular unit of heredity of a living organism is a gene. A gene is a sequence of nucleotides encoding a certain protein according to the principles described below. Apart from the coding sequences, a gene also contains control information as regulatory sequences.

Assumption of the existence of discrete carriers of genetic information (genetic factors) was first suggested by G. J. Mendel in 1865. In 1909, the Danish botanist, W. Johannsen suggested to denote the Mendelian genetic factors by the term "genes." The word "gene" is derived from the Greek word "genesis," which means "birth." In 1911, Th. H. Morgan and his collaborators showed that the gene is a portion of a chromosome, and that a chromosome consists of separate genes located sequentially along its length.

The sizes of genes are all notably different. There are genes that consist of a relatively small number of nucleotides (base pairs), of several dozens, some may comprise more than 2 million nucleotides. The average size is 50,000–100,000 nucleotides.

Various human chromosomes contain different numbers of genes and subsequently of nucleotides. In the largest one, chromosome 1, there are ~250 million base pairs, in the shortest, chromosome 21, about 48 million.

The total number of nucleotides in the human genome is slightly changing while the studies keep going on. Up to July 2014, the total number is equal to 3,095,693,981 [7].

If the number of nucleotides in the chromosomes is determined with accuracy up to a unit, the definition of the number of genes in the human genome is rather problematic. In the early 1960s, when biologists began to estimate the number, they surmised that humans could have as many as 2 million protein-coding genes. Well, the human is thought to be a rather complex creature, perhaps the most complex among other creatures in nature, and that's why it was expected that the highest number of genes are required for it. But as the study went on, the proposed number has dramatically shrunk. By the time the human genome project began in the late 1990s, the highest estimates figured out the number of human genes to be in the range of only 50,000–100,000, and since then the number has been subject to constant revision.

In 2001 the value was decreased to the range of 26,000–30,000. In 2004, the final draft of the human genome reduced the figure to around 24,500, and in 2007 further analysis suggested that it was more like 20,500 genes.

One has to note that here rounded numbers are presented. Even the best recent estimates still have a large amount of uncertainty. The range of estimates has been gradually shrinking, from 50,000–100,000 to some in limits between 18,877 and 22,619 in 2009 [8]. The last value for today is $\sim$19,000. It means that we have fewer genes than nematode worms, which contains about 20,470 [9].

Before the discovery of the DNA structure, it was believed that a gene is a discrete, indivisible unit of heredity. After the discovery, it became obvious that a gene is constructed with many nucleotides, and an elementary particle of genetic material is a nucleotide – a monomeric unit of the polymer molecule of DNA.

In DNA, 1 million "letters" (nucleotides) fit on a line segment $3.4 \cdot 10^5$ nm (0.034 cm) long, having a volume of approximately 10^6 nm^3 (10^{-15} cm^3).

A typical animal cell contains about $3 \cdot 10^9$ nucleotides. This is enough for encoding nearly 3 million proteins. However, calculations based on the estimated number of mutations show [1] that in fact no higher organism can have more than 60,000 vital proteins.

In bacteria, the majority of proteins get encoded by a single continuous DNA sequence. In 1977, it was found that in eukaryotes the coding sequences (exons) intersperse with noncoding sequences (introns). Apparently, this excess amount of DNA was of some importance for evolution.

Of all working genes, at least half are required to sustain any cell's viability, the other half determines to which organ, or tissue, the cell is going to belong. In liver cells, only those genes are activated and only those proteins synthesized that are necessary for the functioning of the liver; similarly, in kidney cells, the only genes activated are those necessary to perform the functions of the kidneys, etc. Thus, all cells contain the same genetic information recorded in the DNA molecules, but as the organism develops, this information in the body tissues or organs is selectively read, which leads to a great variety of cells in the body.

Some genes control the synthesis of proteins, which are not enzymes, and do not create a cell structure. These proteins are combined with other genes so that they are prohibited from the further transformation necessary for the synthesis of other proteins. Such proteins are referred to as inhibitors. Genes producing inhibitor proteins thereby regulate the functioning of other genes, referred to as regulator genes.

It is not yet fully clear how specific genes switch on and off in a cell. Perhaps introns play a certain role in gene expression. Perhaps the control of selective on–off genes is carried out by specific epigenetic mechanisms – that is, those recorded not in the DNA but in some other way, probably associated with proteins that are involved in packaging the chromosomes.

2.1.3 Chromosome Packaging

Although the main genetic information is concentrated in the DNA and the greatest attention was drawn to its properties and functioning, the DNA cannot be considered outside the context of a protein complex, which constitutes the majority of chromosome material.

The enormous amount of genetic information encoded in DNA requires packing the DNA into a small volume. The total linear size of all DNA molecules of a human body is about 1 m (0.34 nm/pair $\times$ $3 \cdot 10^9$ pairs), whereas the diameter of the nucleus of a cell does not exceed 10 μm. Therefore, biological evolution was facing at least two highly complex and in some way mutually exclusive problems. First, the need to pack a DNA into a microscopic volume of a nucleus and, second, for all that, not to lose the ability to extract the necessary information from the DNA, at the right time and in the right combination. Chromatin helps to solve these seemingly contradictory problems.

The concept of chromatin as the stained content of the cell nucleus was introduced into science in 1880, while the term "chromatin" was suggested in 1888.

Normally chromosomes are greatly stretched and therefore not visible in an optical microscope in all phases of cell cycle except mitosis (see Section 2.2.1), since the diameter of a double helix is only 2 nm. In preparation for mitosis, the chromosomes condense and form a helix. First, the DNA is winding up, like a thread, around a certain complex of nuclear proteins (histones). Each molecule makes about two turns on each "spool," and then proceeds to the other one, and so on. The "spool" with the DNA wound around it is called a nucleosome. At this point, the DNA resembles beads on a string. Nucleosomes are arranged along the DNA molecule at intervals of 200 base pairs.

Then the DNA folding process continues with the help of other histone proteins. The diameter of the chromatin (DNA on histones like beads on a string) is 11 nm; the diameter of chromatin fibril (a special way of further DNA packaging) is 30 nm. Further condensation produces a metaphase chromosome, that is, the one placed in a metaphase of the cell cycle (see Section 2.2.1), with one arm's diameter of 700 nm, and of both together at 1400 nm. The length of both arms does not exceed 5 μm. As a result, the chromosome acquires the shape that is usually portrayed by pictures (Figure 2.7 [10]).

This way of chromosome packaging makes it possible to place the chromosomes, with a length of several (up to 10) centimeters each in the expanded state, in a nucleus with a diameter of about 5 μm.

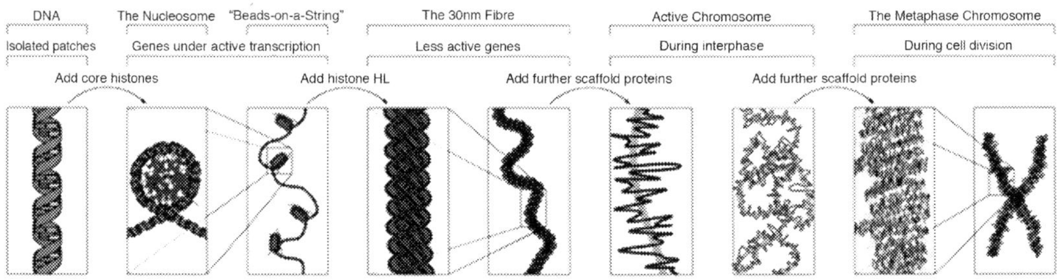

FIGURE 2.7 DNA packaging inside a chromosome. Transferred from [10].

It is important to note that while the DNA is folded, chemicals and free radicals cannot interact with bases. Nucleotide bases, whose interaction with chemicals and products of radiolysis are especially dangerous for the occurrence of mutations, are well protected by the whole protein complex. Only when DNA expands during transcription, replication, or reparation, in the interval between divisions (i.e., during interphase), the bases are available for exposure and hence for mutation. One should bear in mind that during the interphase, the DNA is only partially expanded. For full expansion, there is simply no place in the nucleus. Therefore, DNA bases are open to chemical interaction with foreign substances (i.e., with xenobiotics) only at those moments and only in those parts that are open for replication or transcription. In the double helix, small (1.2 nm) and large (2.2 nm) grooves are distinguished. Some proteins, such as transcription factors, which are attached to specific sequences in double-stranded DNA, typically interact with the edges of the bases in the major groove, where they are more accessible. Obviously, during direct impact of such a destructive factor as ionizing radiation, a chromosome's packaging plays no role because of the radiation's high penetrative power.

2.1.4 RNA, Membranes, Cytoplasm, and Ribosome

2.1.4.1 RNA

Another important cell component is the ribonucleic acid, RNA. Unlike DNA, RNA is single-stranded, ribose taking the place of deoxyribose. Moreover, the place of one of the bases, thymine (T), is taken by uracil (U). Other bases are the same as in the DNA.

RNA performs a variety of functions in a cell. In eukaryotic cells, RNA carries information from the DNA to proteins. The process of information transfer involves several different types of RNA: messenger or matrix (mRNA), ribosomal (rRNA), and transfer (tRNA).

All these types of RNA are synthesized from a DNA during transcription (see Section 2.2.3). Their role in the biosynthesis of proteins (translation process) is different. The messenger RNA contains the information about amino acids' sequence in the protein, ribosomal RNAs serve as the basis for ribosomes, and transfer RNAs (tRNAs) deliver amino acids to the proteins' assembly point – to the ribosome's active centre, "creeping" up the mRNA.

2.1.4.2 Membranes

Apparently, the formation of the outer membrane was one of the key events that led to the emergence of the first cell. Whereas before the membranes evolved, all biochemical processes occurred in a primordial soup, now the future components of the cell received a separate apartment, and selection of RNA molecules, based on the quality of the proteins they encode, became possible.

Through the membrane, the cell receives a variety of substances – the raw material for the biochemical factory, which is what a living cell actually is. Through the membrane, some products also get withdrawn into the extracellular space.

A metabolism exists between the intracellular and extracellular environment. The process by which the required substances get into a cell is called endocytosis. During endocytosis, some parts of the outer membrane are drawn inwards and detach, forming cytoplasmic membrane vesicles. The process of removing substances out of the cell, exocytosis, is a reverse process, in which the intracellular membrane vesicles fuse with the cell membrane, releasing their contents into the surrounding environment.

On the membrane occur a number of important reactions associated with the delivery into the cell of the raw materials for cell metabolism, and the withdrawal of certain products into the extracellular space.

Here it is useful to recall some important concepts. According to its definition, "metabolism" is a set of chemical reactions during which inside a living organism some substances, usually nutritional, transform into other substances, necessary for its vital functions. The used substances then degrade, turn into waste and get withdrawn from the body. Metabolism is split into two phases: "anabolism," the synthesis of complex organic compounds, during which energy is consumed, and "catabolism," a degrading of substances, accompanied by energy release.

A set of chemical reactions involving one substance is called a metabolic pathway. Metabolic processes occur, as a rule, with the participation of enzymes, which regulate metabolic pathways. The main metabolic pathways are the same for many species, starting with the protozoa, indicating the common origin of all living beings. A considerable part of metabolic pathways was studied and described. A summary can be found, for example, in [11].

To maintain the required ratio of the surface area and volume, large eukaryotic cells are forced to increase their surface by bends, folds, and other complications of a membrane shape.

Many components of the cell are also surrounded by their membranes. For example, the membranes surround the lysosomes that contain the stock of enzymes necessary for intracellular digestion. They thus protect the cell proteins and nucleic acids from the action of enzymes. Likewise, membranes surround peroxisomes, where during oxidation of various molecules, dangerous highly reactive peroxides form and decompose.

Basic properties of membranes are determined by properties of lipid molecules – the main components of which membranes are built. Generally, lipids are a broad class of organic compounds – fatty acids that are essential for building membranes. A fatty acid molecule consists of two different components: a long hydrocarbon chain of a hydrophobic (water-insoluble) nature and a carboxyl group, showing hydrophilic (water-soluble) properties. Chemists call this combination of properties amphiphilic or amphipathic. Cell membranes form a specific class of lipids – phospholipids. Inside them, one of the outer chains of higher carboxylic triacylglycerol acids is substituted with a group comprising phosphate. Phospholipids have polar heads and nonpolar tails. Groups forming a polar head are hydrophilic, and nonpolar tail groups are hydrophobic. Two layers of phospholipids connect "tail to tail" to form a lipid bilayer – the structural basis of all cell membranes.

2.1.4.3 Cytoplasm

Cytoplasm is a cell's internal environment surrounded by the membrane. It includes cytosol (hyaloplasm) – a basic transparent substance of the cytoplasm, and its cellular components – organelles, including the nucleus. Cytosol is a colorless, slimy, thick, and transparent colloidal solution. It is here where all the metabolic processes take place. It provides an interconnection between the nucleus and all organelles. The main substance of the cytoplasm is water – 60%–90% of the total weight of the cytoplasm. Bluntly speaking, the cytoplasm is a broth, wherein, dissolved or in insoluble form, a large number of organic and inorganic substances float. It also contains insoluble waste from metabolic processes and spare nutrients.

Cytoplasm viscosity ranges from 2 to 50 cps (the viscosity of water is 1 cps, and glycerol, 1500 cps). It varies in different parts of the cell and at different times of the cell cycle. With a temperature decrease to below 12–15 °C or an increase over 40–50 °C, the cytoplasmic viscosity increases. A characteristic feature of the cytoplasm of eukaryotic cells is its constant movement (cyclosis). It is detected primarily from the movement of cell organelles such as chloroplasts. If cytoplasm movement stops, the cell dies, because it can carry out its functions only in constant motion.

In the cytoplasm, potassium and sodium salts of hydrochloric and carbonic acids and (in smaller amounts) some other salts are dissolved. Obviously, in a cell they are found in a dissociated form. The ion concentration in the intracellular space is maintained with considerable precision at a constant level, and it is different from the concentration of ions outside the cell. These differences in ion concentration play an important role in osmoregulation and signal transmission between cells. Potassium concentration is 139 mM; sodium, 12 mM; chloride ions, 4 mM; and bicarbonate, 12 mM. Outside the cell, the concentration of K^+ is less by approximately one order; that of Na^+ on the contrary is more by an order. The highest gradient between the extracellular and intracellular concentrations exists for Ca^{2+}, whose free ion concentration inside the cell is at least 10,000 times lower than outside.

2.1.4.4 Ribosome

An important process of protein synthesis, in accordance with a program encoded in the DNA, is carried out by a special biochemical machine called "the ribosome." Information on the structure of the protein is carried out of the cell nucleus by the molecules of messenger RNA (mRNA). Molecules of tRNA bring to the ribosome the desired amino acids, and the proteins required for cell functioning are synthesized in the ribosome. The synthesis involves the third type of RNA molecules, ribosomal RNA (rRNA). For more information about one of the most important genetic processes, see Section 2.2.3.

2.2 Genetic Processes

As discussed above, the nucleotide sequence of the DNA molecule contains genetic information. This information has to be stored and made use of. The accuracy and reliability of the storage have to be checked and, if required, this information repository should be repaired. These tasks are performed by genetic processes.

Usually, genetic processes are generally referred to as processes of hereditary information transmission from generation to generation. Of course, these processes play an important, perhaps even essential, role for all living things. But in this book we are going to focus on the transfer of genetic information in somatic cells. In this transfer, the daughter cells after division can perform the same role played by parent cells.

In processing of genetic information, four processes could be pointed out:

- replication – taking a replica, a copy, DNA duplication before each cell division: that is, mitosis;
- transcription – RNA synthesis by DNA template. The process of transcription is followed by a translation process: building proteins from amino acids by RNA template;
- recombination – DNA restructuring, occurring in preparation for meiosis;
- reparation of DNA.

Emergence of a new organism and its subsequent growth occur because of cell division. Even a full-grown healthy organism needs new cells to maintain its normal functioning. Most often, formation of new skin cells, mucous membranes, and blood components occurs. So, the lifetime of intestinal epithelial cells is 1–2 days. Every day, about 70 billion of these cells die. Red blood cells live only 100–120 days, and nearly 2 billion of them die daily. The life span of leukocytes is even less. The cells of the skin's corneous layer continuously exfoliate and drop off.

The body of an adult human every second has to generate several million new cells just in order to maintain its normal condition.

With injury or surgery, the emergence of new cells is especially required to quickly compensate for damage.

Somatic cells' division process is called mitosis.

2.2.1 Mitosis, Meiosis, and Cell Cycle

2.2.1.1 Mitosis

Mitosis is the process for which all genetic mechanisms operate. It is a fundamental process that determines the individual development of an organism from fertilization to death. During mitosis, one parent cell forms two identical daughter cells. Mitotic division ensures the growth of multicellular organisms. Mitosis is a very complex process. Biologists divide it into several stages: prophase, prometaphase, metaphase, anaphase, and telophase. But for more detail, the reader is referred to the special references [1]. Currently, the process of mitosis can be observed as animation on certain sites [12]. In real life, the duration of mitosis in animal cells takes 30–60 minutes, and in plant cells, 2–3 hours. In animated films, it takes just a few minutes.

2.2.1.2 Meiosis

During meiosis (from the Greek "meiosis" [decrease]), or reduction cell division, the double chromosome set of a parent cell is divided in two single sets. The cells with only one chromosome set are called haploid. Thus, generative cells are haploid and normal somatic cells are diploid. Recovery of ploidy (i.e., a transition from haploid phase to diploid) occurs at the confluence of generative cells.

The importance of the processes of cell division for all living beings can be illustrated by the frequently cited statement: "A hen is only an egg's way of making another egg."

Young cells that appeared from division cannot immediately start to divide again. The certain processes of cell growth are required to happen in a cell, along with the recovery of structural components, the necessary synthesis of proteins and DNA molecules, and doubling of chromosomes. All these processes are organized in time and form a particular cell cycle that takes an interval from one division to another.

2.2.1.3 Cell Cycle

The whole complex of processes occurring in a cell from one division to the next and ending with the formation of the two cells of a new generation is called the mitotic cycle or simply cell cycle (or CDC [cell division cycle]). All this cell cycle from one division to another can be roughly split into two phases: the mitosis or cell division phase (M phase) and the interphase. The interphase is a period during which the complex preparation for mitosis occurs in a strict sequence. The interphase occupies not less than 90% of the total cell cycle duration.

More careful analysis allows one to split the interphase into three phases. They are shown in Figure 2.8. Immediately after mitosis, the so-called presynthetic phase G_1 (which means "gap") follows. Then on a certain signal starts the S phase, one of DNA synthesis, in which the transcription and translation take place. The S phase is separated from the mitosis phase by another interval – postsynthetic phase G_2. In the S phase, the major processes of synthesis of DNA occur (which start at the G_1 phase). By the end of the S phase, the chromosomes fully replicate, resulting in a double set of chromosomes within the cell.

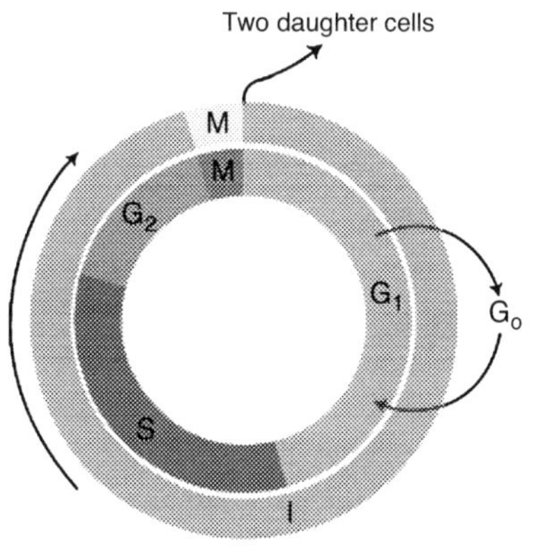

FIGURE 2.8 Cell cycle. The outer circle: I, Interphase; M, Mitosis. The outer circle: M, mitosis; G_1, pre-synthetic phase; S, synthesis phase; G_2, post-synthetic phase; G_0, resting phase.

Duration of cell cycles in various tissues, among different species and at different stages, varies widely. In a frog's early embryo, it can be less than 1 hour; in the cells of intestinal epithelium, it can reach from 10 to 20 hours; in hematopoietic bone marrow cells, it normally takes about 12–24 hours. With a number of other tissues, it may last for several days or even weeks, and in adult human liver, more than a year. Cell cycles in different cells are not connected. Biologists are able to synchronize them during cell division by some artificial methods, like in vitro in a test tube, or rather in a special dish. However, after some time, the simultaneity of the division processes flounders.

In rapidly dividing cells of higher eukaryotes, the M phase lasts from 1 to 2 hours. In principle, all phases of the cell cycle may vary in duration. This refers in particular to the G_1 phase, the duration of which may vary from almost zero to such large values that the cells would seem to stop dividing.

Durations of the G_1 phase can vary under the influence of external factors, but that of the S phase does not change. In G_1, there is a critical moment when the S phase sequence of events starts, and so the external factors have no effect on the further course of the cell cycle. This critical moment is called the starting point. The cells pass through the starting point only after reaching a critical moment.

It is important that the cell cycle has some so-called checkpoints (the late-stage G_1 and the early G_2), and while passing through those, the enzymatic systems can check for DNA damage. If they detect any, the cell's promotion within the cycle gets delayed, the repair processes get activated in order to restore the structure of DNA and to ensure the normal completion of DNA replication and duplication before the cell could enter mitosis. Later we are going to see that the delay in division is a result of the impact of damaging factors on a cell (see Section 4.4.4).

The DNA duplication problem, which creates programs for their functioning in both daughter cells, is solved by the replication process.

2.2.2 Replication

Before the discovery of the DNA structure, the main challenges for genetics were the following: How does a cell store genetic information and how does it pass it over undistorted in the process of division of both somatic and gametal cells, and how that information is implemented in the process of cell development? The solution to the first problem very elegantly follows from the Watson-Crick double-helix model. We are going to return to it later, in describing the mechanism of replication. The second problem is more complicated. Partly, it was solved by the idea of the genetic code, proposed by G. Gamow. Some clarity is going to appear in the description of the mechanism of transcription – translation. This problem at large is called "gene expression," and up to now it has no final solution.

It has been already noted (see Section 2.1.2) that a DNA molecule consists of two polymer chains that are linked together in a double helix by hydrogen bonds between the bases.

During replication, the chains get separated and on each branch grows a new chain, subject to the keeping of the complementarity rule. This means that the nucleotide with the base T joins the base A and the other way round, and the nucleotide with the base C joins the base G and vice versa. Thus, on the two DNA strands, as on a matrix, the two new chains grow and in a cell, two DNA molecules appear instead of one.

In fact, the replication process is very complicated. It is performed by a special multienzyme complex containing the enzyme DNA-polymerase. Polymerase untwines the DNA, and thus two loose chains appear. In Section 2.1.2, the rules of numbering carbon atoms in the DNA molecule are indicated. The chemical bond of nucleotides inside a DNA involves carbon atoms numbered 3' and 5'. The directions of chains in the DNA helix are nonparallel; one has a direction 5'→ 3', and the other, 3'→ 5'.

New DNA chains line up on this chain, like on a matrix. The enzyme DNA-polymerase that controls this process is able to build up a DNA chain only in the direction 5'→3'. A DNA-polymerase that could catalyze the polymerization of nucleotides in the direction 3'→5' has not been detected. Therefore the growth of the "correct" chain is going on naturally. The growth of the second chain also occurs but in short sections (100–200 nucleotides in eukaryotes). These short sections are called the Okasaki fragments. The synthesis goes in the "right" direction of 5'→ 3'; synthesized fragments are then joined into long DNA chains by the same enzyme, which crosslinks breaks in the DNA helix during its reparation, that is, by the DNA-ligase.

New bases match a DNA molecule because of diffusion of the nucleotides already present in complete form inside the nuclear cytoplasm. Various nucleotides may be suitable for that. It is the DNA matrix chain that selects the one that is capable of complementary pairing with the open matrix chain of DNA. The basic material (or substrate as biologists call it) for constructing the new DNA chains are the monomers, molecules of deoxyribonucleoside triphosphate, floating in the cytoplasm. Monophosphate molecules are included in the DNA frame. Therefore, with the chain construction, some chemical reactions occur; deoxyribonucleoside monophosphate joins the chain, and pyrophosphate ($P_2O_7^{3-}$) gets released. At the same time, the hydrolysis of pyrophosphate into inorganic phosphate occurs. This process follows with the release of energy required for the polymerization reaction. Thus, each incoming monomer brings with it energy necessary for its connection to the chain.

A limited replication space that moves along the parental DNA helix is called the replication fork because of its Y-shape. A replication fork is shown in Figure 2.6.

Proteins in the replication fork acting cooperatively form a "replication machine." To make the replication machine move forward together with a replication fork, the entire chromosome in front of it must rotate. For this purpose, there is a special "hinge" (bearing), the DNA topoisomerase. The work of the replication machine, which is the cooperative work of different enzymes, one could see in animation video on the sites [13,14].

During DNA replication, the rate of polymerization fluctuates from 500 nucleotides per second with bacteria to about 50 nucleotides per second in mammals. Obviously, at such a rate, constructing a whole new DNA molecule requires considerable time, which is longer than the duration of the S phase of the cell cycle. Therefore, replication occurs parallelly at many locations of a DNA.

There is one complicating circumstance in the replication process. DNA-polymerase cannot start the synthesis of new polymer chains from the ground up. It can only lengthen an existing chain. In living organisms, this difficulty is overcome with the help of a special enzyme called DNA-primase. This enzyme selects certain sections of DNA that have to initiate replication, and in them it synthesizes initial RNA fragments – the primers. After formation of a primer, the DNA-polymerase continues chain growth with the help of deoxyribonucleotides.

After replication, a cell ends up with a double set of chromosomes, and it is ready to divide. After division, the cell needs construction materials for the growth of the two new cells. The basic material used to build nearly everything in the cell, the one that determines what functions the cell will perform, is proteins. The particular proteins that must be present in this cell, based on hereditary information transmitted from generation to generation, are synthesized in the cell in the process of transcription – translation.

2.2.3 Protein Synthesis

Protein synthesis occurs according to the following scheme:

$$\text{DNA} \xrightarrow[\text{transcription}]{} \text{RNA} \xrightarrow[\text{translation}]{} \text{protein}$$

2.2.3.1 Transcription

The vast majority of the most important functions in a living cell are operated by proteins. In the dry weight of the famous *Escherichia coli* cell, proteins make approximately 50%, while the DNA only 3%.

Proteins form all the enzymes that catalyze essential processes within the cell – the metabolic reactions (cleavage and synthesis of complex molecules), the processes of DNA replication and repair and matrix synthesis of RNA, they regulate transcription, translation, splicing, and other processes. Proteins build the structural cytoskeletal elements collagen and elastin. These are the main components of the intercellular substance of conjunctive tissue (e.g., cartilage). Another structural protein, keratin, forms hair, nails, feathers, and some shells. Proteins carry out motor functions: muscle contraction, cell migration, movement of cilia, and flagella. Receptors that perceive light, sound, impact, taste, and smell are based on proteins. Finally, hormones responsible for signal functions are also proteins.

Necessary proteins are synthesized in a cell in the process of transcription and translation. Transcription is a synthesis of single-stranded RNA molecules on a DNA matrix. A short section of RNA that obtained information about the structure of a certain protein, after some complex processing, gets out of the nucleus and enters the ribosome where on the RNA, like on a matrix, protein is synthesized in the translation process.

The transcription process is performed by a special RNA-polymerase enzyme. During transcription, the RNA-polymerase, moving along the DNA helix, untwists the helix continuously ahead of the point where the polymerization occurs, and twists it again behind this point, releasing the newly synthesized RNA chain. RNA-polymerase begins the transcription from a specific DNA sequence called an origin. This sequence contains the start signal for RNA synthesis. RNA-polymerase moves along the DNA until it encounters on its way another specific sequence of nucleotides, called a stop sign or the point of a transcription termination. Each completed RNA chain is a separate single-stranded polymer molecule, with the number of nucleotides that depends on the type of protein this RNA encodes (from 70 to 10,000 nucleotides). Thus, the so-called messenger or matrix RNA (mRNA) appears.

The section where RNA synthesis takes place is called the reading frame.

Before it gets involved in protein synthesis processes, the RNA must leave the nucleus. For this purpose, while remaining in the nucleus, the RNA undergoes a complex maturating process (so called "processing"), in which some parts of the RNA molecule are removed and others get modified.

Reading speed at 37 °C is approximately 20–30 nucleotides per second. Synthesis of the RNA chain with 5000 nucleotides lasts about 3 minutes.

2.2.3.2 Translation

Proteins are long linear polymers of amino acids. Amino acids have a common feature – they contain a carboxyl group and an amino group bound to the same carbon atom. It is called an alpha-carbon atom. To this atom, a side chain is attached. The proteins of animate nature use 20 different side chains. Amino acids in the proteins are joined "head to tail" using a peptide bond between the carboxyl group of one amino acid and the amino group of another. A peptide bond is a form of covalent interaction. An amino acid bound to the protein chain is called an amino acid residue. Two residues form a dipeptide, and three, a tripeptide, etc. If the number of residues is more than 10, the molecule is a polypeptide, and if that number is more than 40 residues, it is a protein. A typical protein contains several hundred amino acid residues (100–500), and some contain thousands or even tens of thousands. The sequence of amino acids in a protein significantly affects the protein's properties and the function that it performs in the cell and the organism. Apart from the amino acid sequence, the properties of the protein are determined by the way it is rolled into a unique three-dimensional structure.

The order in which amino acids are arranged in the proteins is recorded in a sequence of nucleotides in the DNA. During transcription, this information gets transmitted by the messenger RNA and it carries this information out of the nucleus. In the cytoplasm of the cell, there is a set of transfer RNA – small molecules containing 70–90 nucleotides. Each amino acid has its own tRNA, which may bond to it, with its other end attached to the corresponding place in the mRNA. This allows amino acids to line up in the right order according to the nucleotide sequence in the original DNA.

In a bacterial cell, the duration of one polypeptide chain elongation cycle under optimal conditions is about 1/20 seconds, so that the synthesis of a medium-sized protein consisting of 400 amino acids takes about 20 seconds.

2.2.4 Reparation

For survival, an organism requires very high reliability of genetic information conservation and accuracy of reproduction in each new population of somatic cells. (For the survival of a species, the same is required for the gametal cells.) Section 2.3 describes some possible information distortions due to mutations – damages that arise in a DNA spontaneously, when exposed to damaging factors or as a result of replication errors. To eliminate these distortions in the cells, there is a special DNA repair mechanism.

The two most common types of spontaneous DNA damages are apurinization and deamination.

Apurinization is spontaneous hydrolysis of guanine or adenine due to thermal break of the bonds between purine and deoxyribose. The DNA of each human cell loses about 5000 purine bases per night (adenine and guanine residues).

Deamination is a spontaneous hydrolysis of cytosine into uracil (instead of the group NH_2, cytosine combines with H_2O, followed by tautomeric transition, and thus uracil comes out). The frequency of these transitions reaches 100 per one genome per day.

Damaged areas of the DNA chain are revealed and removed by special enzymes: DNA-repairing nucleases. Different types of DNA damages are revealed by different enzymes. Cells spend a significant portion of their resources in order to produce repair enzymes.

A damaged DNA chain is restored by means of information contained in the complementary intact chain. This is done by another enzyme, a DNA-polymerase. Finally, the enzyme DNA-ligase crosslinks the DNA and thus completes recovery (repair) of the molecule.

Cells can usually cope with single-stranded breaks. Much more serious is the situation when both strands break simultaneously. However, in some cases, even this damage can be repaired by the cell. One scenario of a cell's reaction to the rupture of both DNA strands is a process similar to genetic recombination.

It has to be noted that normally cells have no storage of repair enzymes. Cells synthesize them in response to DNA damage. This may explain why a low level of damaging factors may increase the viability of the cells.

Repair mechanisms solve the problem that was formulated in 1943 by E. Schrödinger [15], before the chemical nature of genes became clear. He noticed that on the one hand, a gene has to be small enough to fit in an atom; on the other hand, it has to be big enough to keep its consistency and not to be the subject to significant changes due to natural fluctuations of the lifetime of many generations. According to the estimates of his time, Schrödinger determines a gene volume as a cube with sides of 30 nm. He wrote, "But let me draw attention at this point to the fact that 30 nm is only about 100 or 150 atomic distances in a liquid or in a solid, so that a gene contains certainly not more than about a million or a few million atoms. That number is much too small … to entail an orderly and lawful behavior according to statistical physics – and that means according to physics" [15]. Now we know how evolution has solved this problem.

2.2.5 Recombination

To make the account more complete, one has also to mention another genetic process – genetic recombination. It allows chromosomes to exchange genetic information, and as a result a new combination of genes forms. This increases the efficiency of natural selection and is important for rapid evolution. In recombination, because of the enzymes' action, the two DNA helixes rupture, exchanging sections, whereupon the spiral's continuity gets restored. The most famous example of recombination is the exchange of segments between homologous chromosomes during meiosis.

2.2.6 Genetic Code

Genetic information recorded in a DNA as a nucleotide sequence is relevant for constructing proteins, which are necessary for the functioning of cells and the whole organism. Proteins are constructed of amino acids. There exists a large number of amino acid varieties, but nature uses only 20 quite specific amino acids. The question arises of how information on the type of required amino acids is encoded in the sequence of nucleotides.

The first to put forward the idea of a genetic code was a well-known American physicist of Russian origin, George Gamow.

There are 20 amino acids and 4 bases. The only way to encode this number of amino acids (given there are 4 bases) is to use 3 bases for each amino acid. Indeed, the number of 3-letter combinations of elements is $4^3 = 64$. Combinations in this case differ from the compounds known from the high school course (arrangements, permutations, and combinations), here each codon may contain the same elements, for example, the codon AAA codes the amino acid called lysine. It is easy to calculate that a doublet code cannot match each amino acid with a code, $4^2 = 16 < 20$. Thus, a genetic code is a triplet; each letter of the genetic code consists of a sequence of three nucleotides. A triplet of nucleotides is called a codon (Figure 2.9). One codon encodes one amino acid. Several hundreds or thousands of codons comprise one gene encoding a single protein.

In other words, the genetic information is encoded in the DNA by a three-letter code, where each letter is one nucleotide and each word is one gene. The overall genetic information stored in the chromosomes of an organism is called the genome.

Because there are more codes than amino acids, so the code is a degenerate one, and several codons correspond to certain acid residues. There are a few meaningless codons that do not match any of the amino acids. Such codons have special functions: they serve as stop-signs, marking the end of a protein chain.

Thus, the code is triplet, degenerate, and nonoverlapping. In addition, the code is universal for the whole of life forms: It is the same for bacteria and primates.

However, let's note an interesting fact. The statement of the code universality has one exception. In cells, there are special organelles, the mitochondria, in which the process

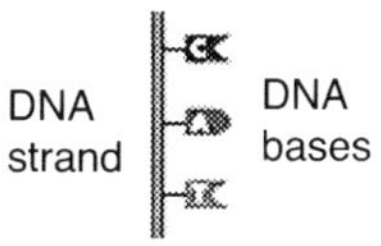

FIGURE 2.9 Codon, three nucleotides. One codon, one amino acid.

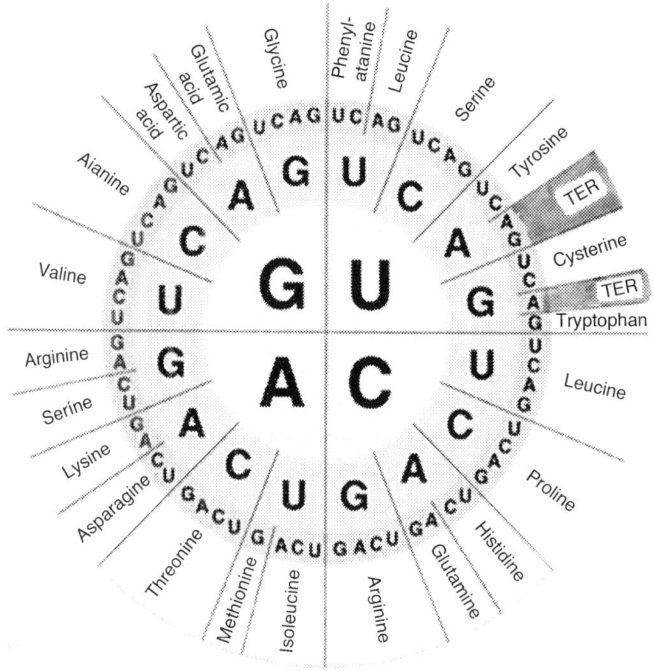

FIGURE 2.10 Genetic code. The first letter of a codon is located in the central circle, the second in the first ring, and the third in the second ring. In the outer ring, the abbreviated names of amino acids are recorded.

of oxidative phosphorylation occurs. Mitochondria have their own DNA, their own RNA-polymerase, their own ribosomes, and their own apparatus of protein synthesis. So, the genetic code of a mitochondria is somewhat different from the code of a nuclear DNA [16].

The genetic code is shown in Figure 2.10. Since the code sets the relationship between nucleotides and amino acids at the translation stage, the picture shows RNA nucleotides rather than a DNA; that is, thymine is replaced by uracil. In the central circle of the table, the first nucleotides of codons are shown; in the next one, the second nucleotides, and then comes the third. On the outer side of the circle are the amino acids. In the genetic code, there are three meaningless codons: UAG, UAA, and UGA, that perform termination function. They correspond to the symbol "TER."

2.3 Abnormalities in the Genetic Apparatus: Mutations

2.3.1 Types of Mutations

Strictly speaking, according to the definition first given by the Dutch botanist and geneticist Hugo de Vries, who rediscovered Mendel's laws in 1901, mutations (from Lat. *mutatio* [change]) are any changes in the genetic apparatus, throughout the genome, in individual chromosomes, their parts, or in a particular gene. Genomic and chromosomal mutations (the latter are called chromosomal aberrations) lead to major alterations in the structure of the genetic apparatus. Gene mutations responsible for violations of the nucleotide sequence in the DNA molecule are most likely to occur. These mutations, the main result of the effects of low doses of radiation and chemicals, will be discussed in this section.

For a physicist, it should be easy to identify classification of mutations with the classification of structural defects of the crystal lattice. Defects can be point, associated with a particular lattice site, or extended, affecting several sites. Generally, point defects of crystal structure are vacancies (i.e., absence of an atom in a lattice site), insertions (i.e., interstitial atoms), and substitutions (an atom in the site or an interstitial atom is replaced by an impurity atom). Thus, if a crystal is made of one kind of atoms, such as a typical semiconductor (e.g., silicon), the range of simple defects is limited to those listed above. If the crystal is more complexly arranged, such as a typical ionic crystal of NaCl, it expands the set of defects. Anion and cation vacancies are two different defects that have significantly different properties. Extended defects are dislocation, block boundaries, cracks, blisters, and other similar defects that disrupt the structure of the crystal much more significantly than point defects.

Similarly, in DNA one can distinguish a group of point mutations. Point mutations are minor; they often affect just one base. This does not mean that they are of little significance. As we will see further, point mutation can significantly affect a cell's function. Point mutations can be of the following types:

Vacancies (deletions): an absence of one or more bases in place.
Substitution: a nucleotide pair substitution. In this case, substitutions may be simple (they are called transitions), when purine is substituted by purine and pyrimidine by pyrimidine, or complex (they are called transversions), when pyrimidine substitutes purine and purine substitutes pyrimidine.
Insertions (or additions): inserting one or more nucleotides at new locations in the DNA molecule. Unlike the similar crystal lattice defects, interstitial atoms, wherein an atom stands not in a regular position but between them, DNA insertion implies creation of a new place in the DNA chain.

These mutations can be illustrated by an alphabetic example. Many references provide these examples using English words (e.g., see [17]). Considering the example, we must remember that first of all, the three-letter codons in the DNA molecule occur in a row, with no gaps, as shown in the first row. Second, the genetic code consists of four letters, A, G, T, and C, and we are forced to use more for illustrative purposes. So,

THEFATCATATETHEFATRAT...: a normal gene
THE FAT CAT ATE THE FAT RAT: a normal gene with distinct codons
THE FAT <u>S</u>AT ATE THE FAT RAT: substitution (S instead of C)
THE FAT CAT THE FAT RAT: deletion of a number of bases, divisible by three (ATE)
THE ATC ATA TET HEF ATR AT...: deletion of a single base (F)

THE FAT <u>DOG</u> CAT ATE THE FAT RAT: insertion of a number of bases, divisible by three (DOG)
THE FAT CA<u>L B</u>TA TET HEF ATR AT…: insertion of two bases (LB).

In these examples, the deletion and insertion mutation types are shown in two variants, when deletion or insertion of the bases is divisible or not divisible by three. Since the genetic code is a triplet, examples with insertion and deletion of bases by a number not divisible by three demonstrate the type of mutation called "frame shift." A shift in codons' partitioning completely changes the content of the phrase, and hence changes at least the part of the protein that comes after the shift.

The literal example shown above makes the phrase, at least in certain "mutations," completely meaningless, unreadable. In DNA, the situation is somewhat different. Among the 64 codon variants (see Section 2.2.6), only three codons are meaningless, that is, have no matching amino acid (UAA, UAG, UGA). These stop protein synthesis. So it may be that the code resulting from a mutation corresponds to some protein, but it is very unlikely that such a protein would be neutral or even beneficial.

Regarding mutations that alter only one codon, there can be different scenarios. Since the code is degenerate, that is, several codons correspond to one amino acid, it may accidentally happen that a codon replacement does not alter the type of amino acid, and therefore does not change the protein. Sometimes the replacement of one amino acid has little effect on the properties of the protein. It is possible that a mutation occurs in a region that does not encode a protein, that is, in an intron. It also does not affect the structure of the protein, although it may affect its expression.

But there are times when this replacement plays a huge role.

Often cited is an example of mutation associated with the replacement of the sixth amino acid in the hemoglobin's beta-chain. There must usually be a Glu (glutamic acid) but as a result of mutation, a Val (valine) takes its place. This mutation is responsible for a severe disease called sickle cell anemia, wherein the oxygen-carrying cells in the blood (red blood cells) change their shape and they pretty much lose their ability to carry oxygen.

Another example of serious biological consequences of replacing one amino acid in a protein synthesized in the cell is described in [18]. Substitution of one amino acid (replacing lysine by glutamate) of 487 amino acids arranged sequentially in a ferment aldehydrogenase is responsible for removing from an organism the acetaldehyde that accumulates during oxidation of ethanol. This leads to an increased sensitivity to alcohol, common in many Asian nations.

Thus, point mutations may have three effects:

a. preservation a codon meaning;
b. changing a codon meaning (missense mutations);
c. forming a meaningless codon (nonsense mutations).

Not only bases can be damaged by various factors, but also sugar-phosphate backbones of the DNA molecule. Rupture of one strand is usually relatively easily fixed by the repair mechanisms of a cell. Rupture of two strands is much harder to repair and leads to very serious consequences.

It is useful to note that in addition to direct mutations discussed above, there may be back mutations – a return to the original order. Further, it will be shown that, for example, the famous Ames test (see Section 5.3.4) is based on reverse mutations.

Mutations in somatic cells have been discussed above. But mutations can also occur in gametal cells. In this case, if the mutations are not removed by a repair mechanism or do not lead to immediate cell destruction, they will be inherited by descendants. In this case, the possible options are the following:

a. These mutations will have no impact on the phenotype. We have already pointed out why some mutations may not affect the life of the cell.
b. Minor harmless changes will occur in the phenotype. One of genetics reference gives an example of such a change: a kitten's ear will be dangling a little bit.
c. Mutations will lead to serious changes in the offspring's phenotype. For example, a single mutation can make an insect insensitive to the action of a well-known pesticide, called DDT. Mutations play crucial role in changing species, but usually they turn out to be lethal and lead to the death of the organism. It is obvious that in this case, such a mutation does not get inherited and will cause no more damage.

2.3.1.1 Mutations in Control Genes

Some sections of DNA encode genes that control other genes, indicating when and where they should start working. Mutations in this part of the genome may have a much more serious impact on the way the organism is formed. Difference in mutations in control and normal genes can be compared to the difference between instructions given to a trumpeter or to the conductor of the orchestra in which this trumpeter plays. Obviously, the instructions to the conductor will have a much more serious impact on the performance than those to just one member of the orchestra. Thus, mutations in control genes may entail a cascade of changes in behavior of the genes controlled by them [18].

Let's note another variant of serious biological consequences of mutations. There are genes that are responsible for control over cells' reproduction, stimulating it to a certain limit and stopping at the phase desired for the organism. Some single changes in such a gene cause it to lose the ability to control the process of cell division, and the gene is transformed into an oncogene, that is, a gene that promotes unlimited reproduction of cells – the cells become cancerous and a cancerous tumor arises.

It's also important to note that mutations are accidental. It is impossible to predict which factor will cause mutation in which gene. Mutations can be beneficial, neutral, or harmful, but they are not related to the needs of the organism. Various factors may affect the frequency of mutations, but they do not affect their orientation. Probability of mutations is unrelated to their usefulness or harmfulness to an organism.

2.3.1.2 Apoptosis as a Kind of Response to Mutation

Apoptosis is the genetically determined process of programmed cell death. Using apoptosis, the body gets rid of unnecessary, "used," or defective cells. Apoptosis is activated by external signals and triggers enzymes that dismantle the cell into parts and take them out into the extracellular space and, further, to the organism's excretory system. Signals that activate apoptosis are confronted by other signals that block apoptosis. Failure to produce the stimulatory signal is essential for unlimited growth of cancerous tumors. By violating production of inhibitory signals, viruses prevent the premature death of the host cell by apoptosis.

In the process of apoptosis, the cell membrane collapses, cytoplasmic and nuclear skeletons break apart, cytosol displaces, chromosomes degrade, and the nucleus gets fragmented. As a result of apoptosis, a cell gets broken down into fragments in 30–120 minutes. Fragments are absorbed by neighboring cells and disappear, which normally takes 24 hours.

Biologists call apoptosis an internal police officer or biological flusher.

2.3.2 Sources of Mutations

The process of mutation is called mutagenesis. There are several sources of mutations. Let us list them. First, spontaneous mutations arise spontaneously as a result of the thermal motion of atoms in a DNA molecule. These mutations account for background mutations, exceeding those that may be caused by other, external factors. External factors that can cause mutations are radiations (ultraviolet and ionizing), chemicals, and viruses.

2.3.2.1 Spontaneous Mutations

Section 2.2.4 identified the two most common types of spontaneous damages of DNA: apurinization and deamination.

The frequency of spontaneous mutations is quite high, but it is almost completely exceeded by reparations, and others. This is the background that we discussed earlier. We cannot separate mutations caused by various factors. So the background mutation is a collection of spontaneous mutations, mutations caused by natural background ionizing radiation and natural background chemicals in the environment. Biologists measure it as one average gametal gene mutation every 200,000 years.

2.3.2.2 Radiation Effect

The destructive effect of ultraviolet (UV) radiation on the structure of DNA is well known. A quantum of UV radiation transfers energy to a nitrogenous base and transfers it to an excited state. Because of the structure of the electron shells of adenine, guanine, and cytosine molecules, the excitation energy gets converted into heat quickly in a nonradiative fashion. In the case of thymine (and only when it adjoins in the chain with other thymine), the excitation energy causes a chemical reaction between adjacent thymine molecules. As a result, a new chemical compound, thymine photodimer, appears somewhere in the

DNA chain. Enzymes that control replication or transcription stop in this place, having met with a strange molecule and, depending on where in the genome it happened, either the further operation of the cell becomes impossible, and thus becomes likely to die, or the mutation persists, and after a few such mutations the cell degenerates into a cancerous one. To eliminate such mutations, evolution developed a special repair system. Interestingly, this system exists in all cells, including those that are never exposed to sunlight, for example, in intestinal epithelial cells.

The effect of ionizing radiation on a cell's genetic apparatus is analyzed in detail in chapter 4.

2.3.2.3 Chemicals Effect

Effect of chemicals on DNA is analyzed in detail in chapter 5.

2.3.2.4 Viruses

First of all, let us recall what viruses are. Viruses (from Lat. *virus* [poison]) are the smallest subcellular particles that can reproduce only inside a living cell. Outside the cell, viruses do not show signs of life and behave as particles of organic polymers. In terms of their size, viruses occupy the place between the smallest bacterial cells and the largest organic molecules – from 0.02 to 0.3 μm. They are not visible in optical microscopes and pass through ultrathin filters. To define them, the term "filterable viruses" was used for a long time. Viruses contain only one type of nucleic acid – either RNA or DNA (all cellular organisms contain both DNA and RNA at the same time). Viruses can be regarded as genetic elements, covered in a protein shell and capable of movement from one cell to another.

For this section, it is significant that viruses can be incorporated into the genome of the host cell, and thereby reprogram its systems. So, with some exceptions, it can be assumed that viruses cause mutations.

2.4 Carcinogenesis

One of the most widespread and dangerous effects of an organism's exposure to either radiation or chemical agents is cancer. High doses of chemicals cause poisoning and radiation – radiation sickness. In both cases, the effect is dose dependent. It is called a deterministic relationship. At low doses, the dose influences not the severity of the effect but its probability; it is called stochastic dependence. And the dominant effect is almost always the same – cancer. Stochastic effects are shown in more detail in Sections 4.4.7 and in Chapter 6. In the meanwhile, let us examine in more detail what exactly is cancer.

Cancer is one of the main causes of death. In the world, there are 24.6 million cases of cancer, and approximately 10.9 million new cases are registered each year. In 2008, 7.6 million people died from cancer, accounting for about 13% of all deaths. About 70% of all cancer deaths occurred in low- and middle-income countries. According to forecasts, the death rate from cancer in the world will continue to grow and in 2030, the number of cancer deaths will exceed 13.1 million [19].

According to recent data for the United States, the number of new cancer cases, averaged by gender, age, and racial specifics, totals 465.2 yearly per 100,000 people for all types of cancer, with a mortality of 178.7 [20].

In Russia, according to the Health Ministry Collegium data for 2002, mortality from cancer is third after cardiovascular diseases and injuries. But, as Russia's Health Ministry experts believe, the reason for only a third place is the lower life expectancy in Russia. Many simply do not survive long enough to contract cancer.

However, perhaps the situation with cancer mortality in Russia is changing. In recent years, cancer mortality has begun to fall. During 11 months of 2011, the mortality rate among cancer patients had decreased by 1% compared to the same period of previous year. If, in 2000, the death rate from cancerous tumors among men in Russia was 200 per 100,000 people, in 2010 this figure dropped to 180.

Reduction in mortality affects not all cancer types but mainly lung cancer, because of an active antismoking campaign, and stomach cancer, because of dietary and lifestyle changes. If this result is not accidental, then we may be talking about a new encouraging trend [21].

Cancer is a collective term that combines a lot of different diseases of various organs and tissues (about 200, different authors suggest different number of types and subtypes of cancer, but obviously there are many). Regardless of the different forms of occurrence of these diseases, treatment methods, and prognosis, they all have one thing in common – the uncontrolled reproduction of cells: out of control of the regulatory and coordination mechanisms in the body. If tissue cells multiply in this way, a tumor (neoplasm) occurs. A tumor is called malignant (cancerous) if its cells are able to invade other tissues, break down barriers, intrude into the bloodstream and lymphatic vessels, and form secondary tumors (metastases) in different parts of the body. Blood cells or hematopoietic tissues may also take a malignant form.

Cancers tumors are classified in accordance with the tissue and cell type from which they originate. Epithelial (epidermal) tumors are called the true cancer or carcinoma (~85% of all cancer cases). Tumors of connective tissue, muscles, and the vasculature system (~1%) are called sarcomas. There are tumors associated with the cells of the nervous system, such as the retinoblastoma (~2%). Cancers of the hematopoietic and immune systems, which do not form tumors, are leukemia and lymphoma (~8%). It should be noted that such quantitative characteristics of danger such as the incidence and mortality rates for various types of cancers may vary considerably. This is because lung and pancreatic cancer are very likely to lead to death, while breast and skin cancer are much less malignant.

According to the most common model, both the main tumor and metastases arise from a single damaged cell, that is, they form a clone. A defective cell transmits its abnormality to its descendants; that is, the damage must be hereditary. In principle, the inherited property may be epigenetic, or manifest itself in changing of a gene expression. However, in the vast majority of cases, hereditary damages are genetic in nature, that is, associated with changes in the structure of the heredity carrier – the DNA molecule. Changes in the structure of DNA and chromosomes, which are fixed and can be transmitted to offspring, are called mutations, and the formation of mutations is called mutagenesis (see Section 2.3).

An important feature of cancer – its strong dependence on age – is discussed in detail in Section 2.5. Also, for cancer, the presence of a latent period between the moment of initiation and the moment when it can be diagnosed is typical.

The main factors that cause cancer are the same as the factors that cause mutations (see Section 2.3.2):

- Spontaneous cancer;
- Viruses;
- Effects of radiation, including ultraviolet;
- Effects of chemicals (carcinogens).

Spontaneous genetic mutations are determined by errors arising from thermal motion of atoms and molecules during DNA replication and repair, as well as some endogenous oxidative processes. The total number of such mutations is immense, but repair systems effectively eliminate many types of damages, and the remaining level is negligible. Frequency of spontaneous mutations varies in different races and nationalities, and depends on the health of individuals and their genetic predisposition. Truly spontaneous mutations are very difficult to separate from mutations caused by unidentifiable background agents that constitute the everyday life. However, some researchers assess the relative role of spontaneous cancer as several percent. The remaining cases of cancer are determined by external factors.

If mutations occurred in a gametal cell and were not corrected by the repair system, they can be passed to the next generation, and the offspring will have the defective gene from birth and all the body's cells will contain this defect. It is not cancer per se, because for a full transformation, cells need a few more impacts. But it is a start. Thus, the rudiments of cancer can be inherited.

Information on the role of viruses in tumor development was obtained at the beginning of the 20th century. In 1910, P. Rous described the first oncogenic virus capable of initiating sarcoma in chickens. In 1945, the viral-genetic theory of tumors occurring under the action of oncogenic viruses was formulated by Russian scientist L. A. Zilber. Although viral carcinogenesis was originally described only in birds and animals, data on the involvement of viruses in the development of some human tumors has been obtained recently. Thus, Epstein-Barr virus causes the development of Burkitt's lymphoma; papilloma virus causes the development of skin and genital cancer, human immunodeficiency virus causes the emergence of sarcomas; and hepatitis B virus is a cause of liver cancer.

Approximately 10% of cancer cases worldwide are caused by live pathogens. This percentage is less in developed countries and more noticeable in developing countries. At least six different viruses and one bacterium, *Helicobacter pylori*, are known to induce malignant neoplasms. Parasites (worms) may also play some minor role.

Apart from those mentioned above, herpes virus, adenovirus, papovavirus, and varicella zoster virus are also related to oncogenic viruses. Typically, these viruses cause infectious diseases, and only one in million causes a malignant transformation.

An opponent of Zilber's virus theory was a prominent Soviet oncology scientist, L. M. Shabad – one of the chemical carcinogenesis theory founders.

In his argument against the universal role of viruses or other exogenous carriers of oncogenic information, Shabad proceeded on the basis that chemical compounds can cause swelling in any organ and every tissue. Then, from the viral-genetic theory's standpoint, this means that all the organs and tissues initially get infected by oncogenic viruses. However, these viruses, without additional (including chemical) impacts, do not cause tumors. Therefore, they cannot be considered as etiological factors in carcinogenesis.

To date, research on the epidemiology of cancer shows that 90%–95% of malignant tumors are caused by carcinogenic environmental factors and lifestyle. Among them, smoking is responsible for 30% of all cancers; nutrition habits, 35%; infectious agents, 10%; occupational carcinogens, 4%–5%; ionizing radiation, 4%–5%; ultraviolet radiation, 2%–3%; consumption of alcoholic beverages, 2%–3%; air pollution, 1%–2%; reproductive factors, 4%–5%; low physical activity level, 4%–5% [22].

Dominant factors determining the risk of cancer in everyday life are chemical carcinogens and radiation. The role of chemical carcinogenesis is variously estimated at between 60% and 90%, and of radiation at between 5% and 25%. These two factors will be described in detail further in the book.

Mutations in cell DNA can occur under the influence of chemicals, carcinogens, which break all barriers built by the body in the way of xenobiotics, intrude into the cell's nucleus, and enter into a chemical reaction with DNA bases. This is called chemical mutagenesis.

Mutations in cell DNA arise either under the direct influence of ionizing radiation or by chemical reactions between the products of cell content radiolysis and DNA. This is called radiation mutagenesis.

Both types of mutagenesis can lead to carcinogenesis.

Transformation of the genome can be caused not only by mutations, but by carcinogen interaction directly with DNA as well – the so-called direct action of the carcinogen. Indirect genotoxic effect is realized by inducing errors in the DNA repair or DNA-replicating enzymes. The probability of direct genotoxic effect is obviously lower than that of indirect.

Modern understanding of the carcinogenesis mechanism [23–25] enables identification of 4 stages of the process: (1) initiation, that is, initial irreversible changes in the cell genome; (2) promotion; (3) progression; and (4) metastasis (invasion). (Some authors use different names for the stages, for example, transformation, activation, and progression.)

The multistage process of carcinogenesis is confirmed, first, by detailed investigation of the cancer dependence on patient age (see Section 2.5); second, by the presence of a long latent period, during which these stages must be realized; and, third, by experimental detection of the above-mentioned stages.

Various factors that could cause cancer differ only in the stage of initiation. Further processes depend very little on the way of initiation. Despite intensive research, all have failed to detect differences in cancers initiated by different factors [26].

In the case of chemical carcinogenesis, it is important to note that most of the carcinogens are only procarcinogens. They acquire an initiating ability (the ability to cause mutations) after metabolic activation – biotransformation of procarcinogens into their reactive forms. By means of enzymatic transformations, carcinogens with very different chemical structures are transformed into reactive metabolites, whose most common feature, as a rule, is electrophilicity. Electrophiles are compounds containing electron-deficient atoms in their molecules. A major role in the metabolic activation processes is played by various enzymes, such as cytochrome P-450 and its isoforms. The electrophilic metabolites resulting from the biotransformation interact with target cells, primarily with their DNA. A variety of products of such interactions has been discovered. These are called adducts.

A certain smaller number of carcinogens are the so-called direct carcinogens, that is, substances that do not require metabolic activation to manifest their carcinogenic effects.

Despite the obvious connection between mutagenesis and carcinogenesis, there is still a possibility of interaction of the carcinogen's molecule with something other than DNA, cellular macromolecules – RNA and proteins. Several carcinogens have the ability to alkylate the tRNA, which in turn violates the translation of mRNA and protein synthesis. But it is the modification of DNA that plays a key role in the molecular processes [27].

A normal cell under the influence of various mutagenesis factors undergoes mutation and gets converted into a precancerous cell. However, in most cases, promotion is required to continue the process. Promoters are substances that are generally not carcinogenic, but their impact is necessary for tumor development. It is believed that the promoters enable expression of the transformed genes. Promoters can be hormones, medicines, or plant waste products, which interact with the cell membrane, receptor structure of the nucleus, or the cytoplasm of initiated cell, and so induce cell division [28]. During promotion, the cell under the action of other factors (promoters, carcinogens) turns into a conditionally cancer cell. It still retains some dependence on the organism, that is, its growth can be blocked by the body's powerful physiological mechanisms, and also has some degree of differentiation. Upon further exposure to the promoter, accumulation of qualitative changes occurs, and a return to the initial state is impossible even with termination of the promotion. Experiments show a synergistic interaction between the initiating factor and the promoter for many different organs and cell systems. In particular, it has been shown that promotion reduces the latency period. It is generally accepted that the same is true if the initiating agent is radiation.

Several carcinogens are both initiators and promoters of tumor growth. In this case, they are called complete carcinogens.

The next stage (progression) leads to transformation of an already existing, but benign, tumor into a malignant state. New developments in both genotype and gene expression enable unlimited growth of the tumor cells, which start suppressing less aggressive cells. It is shown that radiation can enhance the progression process.

The currently accepted theory of carcinogenesis suggests that mutations are an indispensable prerequisite for cancer development. After mutations occur, many of the organism's defense mechanisms can prevent further conversion of cells and cancer can be avoided, but without mutations, there can be no cancer. It would seem there should be a clear link between mutagenesis and carcinogenesis. However, the situation is more complex, the relationship between these phenomena is not simple. Analysis of databases on carcinogenic and mutagenic properties of various substances shows that not all carcinogens exhibit mutagenic activity and not all mutagens are carcinogenic.

Possible lack of mutagenic properties in carcinogens may be due to an indirect action of the carcinogen on genotype, as mentioned previously. But the lack of carcinogenic activity in substances, in which mutagenic properties are reliably identified in bacterial tests, can cause confusion. Does it ruin one of the main paradigms of carcinogenesis? It is a serious question, because if mutagenicity tests cannot identify carcinogenic substances, they should be excluded from test practice. This issue is being actively discussed in the scientific literature. Biologists provide the following answers to this question [29,30].

In bacterial mutagenicity tests, a potentially dangerous chemical gets into the cell passing only one barrier – the cell membrane. In animal tests, which may reveal carcinogenic substances, they must go through numerous barriers that nature has built in the way of xenobiotics. Which specific barriers the substances pass through depends on the method of administration. Substances tested are administered to an animal with food or drink, applied to the skin or inhaled. The first barriers for a substance administered with food are saliva and gastric juices, which can degrade potential carcinogens. Significant barriers to xenobiotics are the mucous membranes of the gastrointestinal and respiratory tracts. As a result of passing through these barriers, a substance gets deactivated before it reaches the target organ cells.

Control of the processes in the cell, in particular related to division and proliferation (gene expression), is monitored by the positive and negative regulatory genes. Genes that stimulate cell transformation into a cancerous one as a result of impacts are called proto-oncogenes. Under the influence of damaging factors, they become oncogenes. Genes that encode proteins related to the negative growth regulators are called suppressor genes or anti-oncogenes. Activation of proto-oncogenes occurs in a variety of mechanisms, some of which have been studied rather well [25].

It is important that to induce cancer cells, the above-mentioned DNA damages must occur in descendants of the target cells. It can only happen to those cells that will divide, from which the well-known Bergonie-Tribondeau rule clearly follows. In 1906, these French scientists formulated the rule, according to which sensitivity of cells to radiation is directly dependent on their ability to divide at a given point in time. This means that the ionizing radiation is particularly dangerous for a growing organism, and that the younger the body the higher the radiosensitivity. This rule was formulated for radiation damages, but is apparently true for any means to influence cell genetic apparatus: "all that divides" get primarily affected. One possible role of promoters is to stimulate cell division. It is known that cancer development gets promoted by various mechanical damages, irritations, or wounds because their healing is associated with active cell division. In principle, DNA damage can be recognized by regulatory and protective systems of the body, supporting the molecular cellular homeostasis, and then eliminated by activation of various repair systems.

The probability of a certain mutation in a certain gene may be very small. But after division the population of cells that are already partially defective, but have retained their ability to divide, can grow to a volume of $\sim 10^6$ cells. Then, even rare events for singular divisions can occur in the entire population with sufficient probability.

Discovery of oncogenes and anti-oncogenes enabled formulation of the modern multi-stage model of cancer. Moreover, the proposed model is quite versatile and can be applied to a large number of cancers, perhaps even to all of them [31]. According to this model, the transformation of a normal cell into a cancerous one is due to six major changes in cell physiology that jointly enable tumor growth:

1. Independence of growth signals: In order to divide, a normal cell requires a special signal to change from sleep mode to an active reproduction stage (see Section 2.2.1). According to current data, none of the normal types of cells can reproduce without such a signal.
2. Insensitivity to signals stopping the growth: Normal tissues are affected by signals that inhibit cell reproduction and establish cell rest (sleep) and homeostasis. Antigrowth signals can transfer a cell into resting state G_0 or keep it in the G_1 phase of the cell cycle (see Section 2.2.1).
3. Avoidance of programmed cell death (apoptosis; see Section 2.3.1): The ability of tumor cells to increase in number is determined not only by the rate of cell division but also by the speed of cell deterioration. Programmed cell death – apoptosis – is the main source of deterioration.
4. Unlimited reproduction potential: In principle, the three above-mentioned cell malfunctions seem to be sufficient for unrestrained cell growth and tumor development. However, recent studies show that destruction of this intercellular signal system does not yet provide tumor growth. Many, perhaps all, types of mammalian cells have an internal, autonomous program that limits the growth in their numbers. This program should also be destroyed so that a cell clone can grow indefinitely to macroscopic size, threatening the life of the organism. It is proven that cells in an vitro culture have limited reproduction capacity [32]. After a community of cells undergoes a certain number of divisions, the process stops.
5. Strong angiogenesis: Supply of cells with oxygen and nutrients is essential for their normal functioning. That is why all cells of any tissue are located within 100 μm from the nearest capillaries.
6. Ability to invade other tissues and create metastases: Cancer cells are able to leave the tumor where they've grown, get transported through the bloodstream or lymphatic system to the new location, establish themselves there, and start growing at the new place where food and space are not limited (at least in the beginning). This ability is an extremely important element of cancer because it is metastases that make a tumor particularly malignant. They cause more than 90% of deaths due to cancer.

The need for such a major coincidence of changes in the cell or its descendants may explain why cancer is relatively rare during a human's life.

The order in which the listed disruptions in normal cell functioning must occur can vary significantly for different types of cancer. Furthermore, in some tumors, particular genetic damages can occur simultaneously, thereby reducing the number of distinct mutational steps required to complete tumor development. In other tumors, the necessary damages require two or more mutations. So that in total, five to eight mutations may be required. As shown in the next section, it is well correlated with the slope of the curves of dependence of cancer induction probability on age.

2.5 Cancer and Age

The dependence of cancer on age is an important feature of cancer. This relationship is also important for understanding the mechanism of cancer occurrence.

When analyzing the age dependence of cancer or cancer mortality, it is useful to compare a cancer cell with a radioactive nucleus. It is known that a radioactive nucleus has no history. Its chances to decay don't depend on its age. The well-known law of radioactive decay is based on this particular condition.

A cancer cell, on the contrary, has a history. For a normal cell to become a cancer one, certain events within it, that is gradually accumulating mutations, are needed. Because of this, the probability for a normal cell to become a cancerous one increases with age. Nearly 60 years ago, the first mathematical models appeared describing the dependence of cancer on age. The first one was the work of Armitage and Doll [33]. The authors' reasons were as follows: if cancer is caused by just one mutation, the probability of its occurrence should be the same for different age groups. In this case, the situation is similar to radioactive decay. If a normal cell's transformation requires two mutations and the mutation probability is evenly distributed over time, the incidence of cancer is going to be proportional to the age. In the case of three mutations, one can expect a quadratic dependence of this frequency on age; in the case of four mutations, a cubic dependence is therefore expected, etc.

On this basis, the authors proposed the following formula for cancer incidence at age t:

$$I(t)=ct^{k-1}, \tag{2.1}$$

where $I(t)$ is the number of cancer occasions per 100,000 people per year in a relevant age group, c a parameter characterizing general susceptibility to cancer, and k the number of stages in cancer development. Other formulae were also suggested, but this one fits better the statistics in a certain age range (~25–70 years).

Cancer statistics analysis showed that, indeed, in double logarithmic coordinates, the dependence is expressed by the straight line, as it should be if the dependence expressed by the formula (2.1) is valid.

Figure 2.11 shows in the double logarithmic scale the dependence of the death cases on a patient's age for various types of cancer. The graph has been built by the author on the

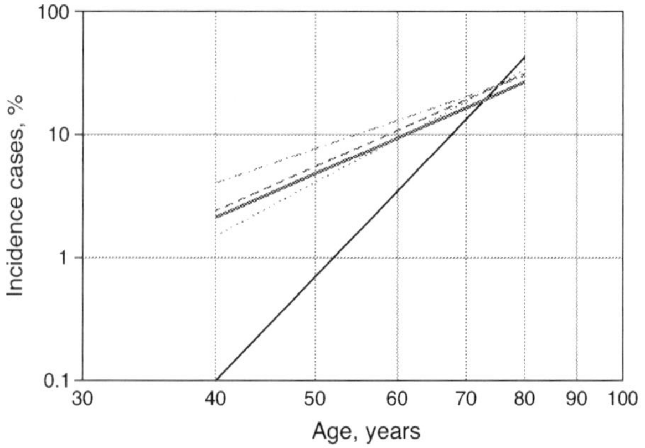

FIGURE 2.11 Dependence of the death cases on patient's age for various types of cancer. Black: prostate; dotted line (blue dotted line in the web version): lung and bronchus; thick line (green line in the web version): esophagus; dashed line (red dotted line in the web version): colon; dash-dotted line (orange line in the web version): stomach. Double logarithmic scale. The data are limited by the young (>40) and old (<80) ages. On the basis of data from SEER [41].

basis of data from the Surveillance, Epidemiology, and End Results (SEER) program [34]. It can be seen that in this scale the dependence, indeed, can fit the straight line if data are limited by the young (<40) and the old (>80) ages. The lines are normalized to the same number of cases, but the real number of cases is different, for example, the annual number of deaths for lung and bronchus cancer is ~345,000 and for stomach cancer ~58,000.

The graph in Figure 2.11 shows just a small part of the data, but they are all similar. The decline of the curves is slightly different but in most cases it is near the sixth power of age. At the same time, despite some variation of the parameter k, it is constant for a given tissue. The strong difference is observed for prostate cancer, as shown in Figure 2.11. The curve for prostate cancer has a slope of about 10. The value $k \sim 10$ exceeds a reasonable range of possible values (see Section 2.4). To solve the problem, Cook and colleagues [35] suggested that the prostate is an organ that appears and begins to function not immediately after its birth, but with some delay t_0. In addition, prostate cancer has a quite long latency period, that is, the period between the initiation and the time when the tumor can be diagnosed. In this case, the frequency of cancer occurrence should follow the relation

$$I(t) = c(t - t_0)^{k-1}. \tag{2.2}$$

The fitting of real data showed that the best result for all groups is obtained at $t_0 = 32.5$ years. The slope of the curve decreases and becomes $k \sim 5$. Besides, the dramatic improvement in prostate cancer diagnostics, following the introduction of a test for prostate-specific antigen (PSA) in 1991, did not clarify the situation.

The probability of many diseases increases with age, but for cancer this dependence is especially high. It is also worth mentioning that with women, cancer of specifically female organs slows down after 45 years, and cancers of other organs, the frequency of which increases in proportion to the sixth power of age before and after the menopause decade, significantly slows its growth in this period.

Child and adolescent cancers are very rare, but still much more common than suggested by extrapolation of the sixth power dependence to smaller ages. That's why they are excluded from the data for Figure 2.11. An unexpectedly high frequency of child cancers can be due to the high rate of cell division in the prenatal and infant period.

Simple extrapolation of the model for the age of more than 70 years (statistics were reliable only up to this age) leads to the formulation of the thesis "If a person lives long enough, sooner or later he is going to have cancer," or in other terms, "everyone should get cancer, but not all live long enough to have it".

However, even in the early studies, in most cases the points for older ages are below the fitting line. Nevertheless, the data was insufficient for the major conclusions. In early works [33,36], only the data for the age of ~70 years are presented, since the data on diseases for the older ages were considered unreliable.

The fact is that there were certain difficulties with determining the cause of death of the elderly people. A cystoscopy would be rarely carried out for that age. The physicians tended to indicate just the old age as the cause of death, rather than to conduct a more thorough analysis and to provide a more accurate diagnosis. Moreover, the increasing ability of physicians to cure cancer also reduces mortality.

But in time, the accumulation of statistical information has been improved. First, data on the disease incidence in elderly people became more reliable. Second, statistical studies have turned from mortality fixation to disease fixation. As a result, analysis of cancer dependency on age could be also made reliable for the older age groups. A detailed analysis of the situation in older age groups was given in the article [37].

A typical curve of cancer dependence on age considering the statistics on elderly people (>75 years) is shown in Figure 2.12. The graph has been built by the author on the basis of the data of SEER [34]. Practically all published data show that the decline of cancer incidence with age practically for all types of cancer can be observed very clearly. The data are similar in different groups of population and vary only on the height of the curve. There is an obvious remarkable uniformity for the age that matches the curve peaks, despite the fact that the frequency of different cancers may differ by more than 100 times. So, the maximum frequency of prostate cancer with men in the analyzed selection is ~1200 cases per 100,000 people, but, for example, the frequency of malignant granuloma (Hodgkin disease) is only 5 cases.

Also noteworthy are the different cancer frequencies with men and women. With women, it is lower for almost all types of cancer. For example, laryngeal cancer with men at the peak is ~40 cases per 100,000 people in a year, and with women ~5.

New data revealed with certainty a paradox: at older age, the curve of cancer by age slowed its growth, becoming flatter, passing through a maximum somewhere around age 80 years and then begins to fall. Similar behavior was shown by curves for different types of cancer for both men and women. The collected data show that practically in all cases the probability of cancer decreases from approximately 80 years of age, and the risk of disease disappears at the age of about 100.

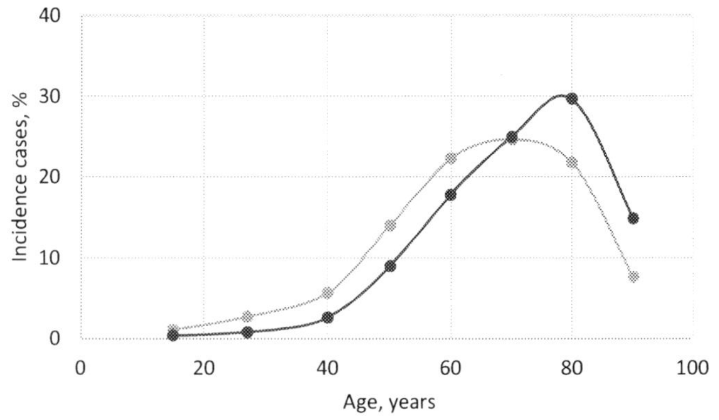

FIGURE 2.12 Age distribution of cancer diagnosis (light gray line (blue line in the web version)) and death (dark gray line (red line in the web version)) averaged for all sites, both sexes and all populations. On the basis of data from SEER [41].

Although of course, some cases have their own peculiarities. For example, in the case of testicular cancer, the maximum corresponds to the age of 25–40 years and by 80 years of age the probability of cancer decreases by a factor of hundreds, practically down to zero.

In order to describe cancer's dependence on age, Pompei and Wilson with colleagues [37] proposed to introduce an additional term into the expression (2.1)

$$I(t)=ct^{k-1}(1-bt), \tag{2.3}$$

where c and t have the same meaning as in (2.1), and b is a parameter, which meaning is best understood from its inverse value $1/b$, showing the age at which $I(t)$ equals zero. Expression (2.3) is a special form of the beta function. Factor $(1 - bt)$ is interpreted as a reduction of the probability of the last stages of cancer at older age.

The incidence of cancer does not just decline after the maximum, but reaches zero at the age around 100 ($1/b \sim 100$ years). And the range of $1/b$ values, obtained by fitting the statistical data by the function (2.3) for different types of cancer (for different organs and tissues), was very small. This made it possible to change the long-dominant theory of universal doom to cancer. New data suggest that "if a person lives long enough, there is a chance to avoid cancer altogether."

It turned out that the cancer inhibition with aging is observed not only in humans but also in experimental animals – mice in particular. Usually when testing carcinogens on rodents, they have to be slaughtered at the age of approximately 2 years. But if they are let to live their entire normal life span (about 3 years), their cancer mortality decreases with age. In one of the studies [38], it was discovered that with mice the incidence of cancer peaks at approximately 800 days of age, and then falls.

The reduced probability of cancer detection with the elderly people is now quite evident, but it is not completely clear whether it is a biological effect or some phenomenon associated with the peculiarities of cancer diagnostics.

Here one has to mention some possible complicating effects that could affect the course of the discussed dependencies. Of course, the first and seemingly obvious reason may be some errors in diagnosis for the older ages and a smaller sample size than for the younger ages. Another possible reason is the existence of yet unknown biological concepts of predisposition to cancer. If there are both normal and predisposed people in the younger population groups, then in time the predisposed ones get sick and die, and in the subsequent age groups, the number of the predisposed people gets smaller and, therefore, the likelihood of disease decreases. A third explanation is the assumption that the multistage model of cancer, resulting in dependency of cancer incidence on age by type (2.1), assumes a constancy of mutation probability in time. However, it is obviously not so. Older people may change their diet and thus exclude certain food carcinogens. Older people can lose weight, which is important for certain cancer types. They can reduce alcohol and tobacco consumption, or switch from plain to filter cigarettes. On retiring, older people can stop being exposed to carcinogens with which they dealt occupationally, etc.

The detailed analysis of the above-mentioned and other similar reasons showed that they are uncertain and unlikely [34,39–41]. But there is still no explicit explanation of decrease of the probability of cancer with the elderly people. But it is, hopefully, not the result of all sorts of side effects, some of which we have just described, but some biological effect of suppressing cancer cells in time. It is suggested that precancerous cells can be destroyed or deactivated at a speed that exceeds the speed of their creation. This relatively new thesis requires further study.

References

[1] Alberts B, Johnson A, Lewis J, Raff M, Roberts K, Walter P. Molecular Biology of the Cell. 5th edition. New York, NY: Garland Publishing Inc.; 2007. p. 1392.

[2] Elliott WH, Elliott DC. Biochemistry and Molecular Biology. New York, NY: Oxford University Press; 1997.

[3] Parker ET, Cleaves HJ, Dworkin JP, Glavin DP, Callahan M, Aubrey A, Lazcano A, Bada JL. "Primordial synthesis of amines and amino acids in a 1958 Miller H_2S-rich spark discharge experiment". Proc. Natl. Acad. Sci. USA 2011;108(12):5526–31.

[4] Dokins R. The Selfish Gene. New York, NY: Oxford University Press; 1976. p. 224.

[5] Cell (biology), the article in Wikipedia. http://en.wikipedia.org/wiki/Cell_(biology), the picture is from the site: http://en.wikipedia.org/wiki/Cell_%28biology%29#mediaviewer/File:Celltypes.svg.

[6] Sverdlov ED. DNA in a Cell: From a Molecular Icon to the Problem of "what is Life?". Vestnik Russian Acad Sci 2003;73(6):497–504. In Russian.

[7] The article in Wikipedia Human genome. http://en.wikipedia.org/wiki/Human_genome.

[8] Pertea M, Salzberg SL. Between a chicken and a grape: estimating the number of human genes. Genome Biol. 2010; 11(5): 206. Published online May 5, 2010. http://www.ncbi.nlm.nih.gov/pmc/articles/PMC2898077/.

[9] The Physics arXiv Blog on Jan 3. https://medium.com/the-physics-arxiv-blog/human-genome-shrinks-to-only-19-000-genes-21e2d4d5017e.

[10] The figure is taken from the site in Wikipedia: http://commons.wikimedia.org/wiki/File:Chromatin_Structures.png. The author is Richard Wheeler, his user page http://en.wikipedia.org/wiki/User:Zephyris.

[11] Salway JW. Metabolism at a Glance. 3rd edition. Blackwell Publ.; 2004.

[12] The animation of genetic processes, in particular, mitosis in YouTube:. http://www.youtube.com/watch?v=VUEFQxZab4Q – Mitosis 1 min 26 s;. http://www.youtube.com/watch?v=OlbNsA1ueS0&feature=related – Inner life of a cell – 8 min;. http://www.youtube.com/watch?v=aZd9DZIdt5Y&feature=fvwrel – DNA, RNA, ribosome – 23 min. 22 s;. http://www.youtube.com/watch?v=E-R-I7imPCE&feature=related – DNA replication – 2 min;. http://www.youtube.com/watch?v=yKW4F0Nu-UY&feature=related – Inner life of a cell – 8 min.

[13] McGraw-Hill Higher Education. http://highered.mheducation.com/olcweb/cgi/pluginpop.cgi?it=swf::535::535::/sites/dl/free/0072437316/120076/micro04.swf::DNA+Replication+Fork.

[14] Harvard University Information Technology. http://sites.fas.harvard.edu/~biotext/animations/replication1.swf.

[15] Schrödinger E. What is Life? The Physical Aspect of the Living Cell 1955.

[16] Frank-Kamenetzky M. The main molecule. Moscow: Nauka, library Quantum; 1983. p. 159 (in Russian: М.Д. Франк-Каменецкий. Самая главная молекула. М.: Наука, 1983, 159 с.).

[17] Learn Genetics. Genetic Science Center. How Do Cells Read Genes? http://learn.genetics.utah.edu/content/molecules/dnacodes/.

[18] Knorre DG. Biochemistry of nuclear acids. Soros Education J 1996;(3):11–6. In Russian.

[19] World Health Organization. Media centre. Cancer. Fact sheet N°297. Updated February 2014. http://www.who.int/mediacentre/factsheets/fs297/en/.

[20] National Cancer Institute. Surveillance, Epidemiology and End Results. Annual Report to the Nation on the Status of Cancer, 1975-2008. http://seer.cancer.gov/report_to_nation.

[21] RIA News. The cancer mortality in Russia has declined over last ten years. The talk of D.G. Zaridze, the president of the Russian anticancer society. http://ria.ru/society/20120202/554792903.html (in Russian).

[22] Zaridze DG. Epidemiology, mechanisms of carcinogenesis and cancer prophylaxis. The talk on the III Congress of Oncologists and Radiologists of CIS in Minsk, 2004. http://health-ua.com/articles/989.html – in Russian.

[23] UNSCEAR 2000 Report Annex F. "DNA Repair and mutagenesis." http://www.unscear.org/unscear/publications/2000_2.html.

[24] UNSCEAR 2000 Report Annex G «Biological Effects at Low Radiation Doses." http://www.unscear.org/unscear/publications/2000_2.html.

[25] Khudoley VV. Carcinogens. Sankt Petersburg 1999;419. In Russian.

[26] BEIR VII PHASE 2, 2006. http://www.nap.edu/openbook.php?record_id = 11340&page = R1.

[27] Jones P. DNA methylation and cancer // Oncogene 2002;21:5358–60.

[28] Kutsenko SA. Basics of Toxicology. Saint Petersburg, 2002. Chapter 6.3. Chemical carcinogenesis – in Russian. http://www.medline.ru/monograf/toxicology/p6-specialformsoftoxpro/p3.shtml.

[29] Zeiger E. Mutagens that are not carcinogens: faulty theory or faulty tests? Mutation Res 2001;492:29–38.

[30] Zeiger E. An interesting state of affairs in genetic toxicology. Environ. Mol. Mutagen 2000;82:82–5.

[31] Hanahan D, Weinberg R. The Hallmarks of Cancer. Cell 2000;100:57–70.

[32] Hayflick L. Mortality and immortality at the cellular level: a review. Biochemistry 1997;62:1180–90.

[33] Armitage P, Doll R. The age distribution of cancer and a multi-stage theory of carcinogenesis. Br. J. Cancer 1954;8:1–12. reprinted 91 (2004) 1983-1989.

[34] National Cancer Institute. SEER Cancer Statistics Review. Age Distribution (%) of Incidence Cases by site, 2003-2007 for 17 areas in the USA. http://seer.cancer.gov/archive/csr/1975_2007/results_merged/topic_age_dist.pdf.

[35] Cook PJ, Doll R, Fellingham SA. A mathematical model for the age distribution of cancer in man. Int. J. Cancer 1969;4:93–112.

[36] Nordling CO. A new theory on the cancer-inducing mechanism. Brit. J. Cancer 1953;7:68.

[37] Pompei F, Wilson R. Age distribution of cancer: the influence turnover at old age. Human Ecol. Risk Assess. 2001;6:1619–50.

[38] Pompei F, Polkanov M, Wilson R. Age distribution of cancer in mice: the incidence turnover at old age. Toxicol. Indust. Health 2001;17:7–16.

[39] Harding C, Pompei F, Lee EE, Wilson R. Cancer suppression at old age. Cancer Res. 2008;68:4465–78.

[40] Mdzinarishvili T, Gleason MX, Kinarsky L, Sherman S. A generalized beta model for the age distribution of cancers: application to pancreatic and kidney cancer. Cancer Inform. 2009;7:183–97.

[41] Olschwang S, Yu K, Lassert C, Baert-Desurmont S, Buisine M-P, Wang Q, et al. Age-dependent cancer risk is not different in between *MSH2* and *MLH1* mutation carriers. J. Cancer Epid. v. 2009, article ID 791754, 6 pages.

3

Evaluation of the Action of Hazardous Factors on a Human

At high doses of adverse factors, a human develops a disease, the severity of which increases with increasing dose. Typically, some diseases have a dose exposure threshold. Such diseases are called deterministic or threshold. At doses considerably smaller than the threshold, no clinically definable damage occurs, but there could occur after a considerable time a manifestation of adverse effects – long-term effects (see Section 4.4.7). A characteristic feature of long-term effects is the probabilistic nature of the upcoming damage. Such effects are called stochastic. Severity of the consequences arising does not depend on the dose; it is the effect's probability that is dose dependent and there is, in most cases, one consequence – cancer (for other possible consequences, see Section 4.4.7).

The stochastic nature of the lesion reveals itself in the fact that some people exposed to carcinogens or ionizing radiation have cancer but others do not. Cancer can sometimes occur in people not exposed to such an influence. Thus, cancer is not a mandatory consequence of exposure and the effect is not necessary for cancer. Association between exposure to a carcinogen and the cancer is probabilistic or statistical. Cancer risk is dose dependent, rather than the severity of the lesion. Therefore, the result of exposure of any selected group of people to carcinogens (of all kinds, including radiation) is expressed in random numerical values.

Quantification of this impact usually involves the extensive use of a branch of science called "biostatistics." Those interested in this topic should refer to the guidelines [1–8]. A more precise name for this direction is "clinical epidemiology" [1], or more generally, "evidence-based medicine" [9,10].

It's assumed that the readers are familiar with the basics of experimental data-processing methods. In this chapter, just the following few aspects of biostatistics required for further comprehensive reading of the book are considered:

3.1 Calculating risks.
3.2 Verification of tests.
3.3 Probit analysis.

3.1 Calculating Risks

Biologists use the concept of risk while performing statistical analysis on adverse factors' impact on a body. The concept of risk includes both the direct threat from the alleged event and a quantitative measure of the probability of such an event. Risk is the universal measure of unpleasant consequences and is widely used in many fields of human activity.

There are risks of natural disasters, from technical human activity, the risk of financial transactions and business activities, etc. [11,12]. In this section, we will get acquainted with the particulars of calculating risk as a quantitative measure of the impact of a couple of damaging factors – radiation and chemical carcinogens – on the body.

The most reliable results could be obtained in experimental studies. An experimenter deliberately creates a group of subjects, and a control group, sets the exposure dose, develops a research protocol, and takes steps to minimize the role of any additional factors. But experimental studies are possible only on animals.

Another option to study the influence of factors affecting the health is the so-called method of intervention. As one cannot work with people by introducing damaging factors, so it is possible, rather, to remove the factor. For example, a group of volunteers may exclude a food product suspected to be harmful, and then one can compare the results of such exclusion between the study and control groups. Processing of the results of these studies is similar to those that are going to be discussed in this section.

However, in reality, most information can be obtained from observational epidemiologic studies. They are called observational because the role of the researcher is to monitor the influence on people of adverse factors that have already occurred, and their consequences.

The main problem of observational studies is that the observed group of patients may differ not only by the fact that some are affected while others are not but also by a large number of other factors such as lifestyle, socioeconomic parameters, health status, attitude to work, smoking, alcohol, and many other almost immeasurable factors. Therefore, the role of the investigated factor is quite difficult to identify.

Biologists use the term *cohort* to identify a group of people united by some property. Originally, a cohort meant a unit of the Roman army numbering about 500 people; 10 cohorts made up a legion. Cohort as a name in epidemiologic studies is quite reasonable really: a group of people, "marching" from exposure, to any manifestation of the impact of that exposure, from the effect to the disease [13,14].

Epidemiologists organize observational studies in several different ways with varying designs. Here, we will discuss the two most common ways: cohort studies and case–control studies.

A case–control first identifies a diseased individual (the case) and remaining healthy (control) individuals. Then, in retrospect, their exposure to damaging factors (radiation dose or chemical carcinogen) is determined.

In cohort studies, the assessment begins with an explanation of the exposure, and subsequently the ill and healthy are revealed. In accordance with operating procedures, cohort studies may be retrospective, when the situation with exposure that has already occurred, and diseases revealed, is analyzed, or prospective, when the researcher starts the analysis either from the beginning of exposure, or during the process, but even before the onset of overt disease.

In both cases, quantitative risk indicators can be computed on the basis of results obtained in the observations.

Let us introduce the notion of the incidence rate $\lambda(t)$, which is the ratio of the number of new disease cases over a certain period of time to the total number of persons healthy at the start of this period. Obviously, the incidence of the disease depends on the duration of the interval and from its position in time, especially in the interval between an exposure to a carcinogen and the research. Note that the time integral of the frequency of the disease is called the cumulative frequency.

$$\Lambda(t) = \int_0^t \lambda(u)du.$$

Let $\lambda_E(t)$ and $\lambda_U(t)$ be the incidence of people exposed and unexposed to effects, respectively. From these values, three parameters can be constructed:

a. the difference $\lambda_E(t) - \lambda_U(t) = \mathrm{EAR}(t)$ (This difference is called the excess absolute risk [EAR].)
b. the ratio $\lambda_E(t) / \lambda_U(t) = \mathrm{RR}(t)$ (This ratio is called relative risk [RR].)
c. the ratio $\mathrm{ERR}(t) = [\lambda_E(t) - \lambda_U(t)] / \lambda_U(t) = \mathrm{RR}(t) - 1$ (This value is called the excess relative risk [ERR].)

Rearranging the members shows that EAR(t) describes the incremental increase in the incidence over the spontaneous disease incidence in people who are not exposed to destructive factor $\lambda_U(t)$, and RR(t), a multiplicative increase.

$$\lambda_E(t) = \lambda_U(t) + \mathrm{EAR}(t), \tag{3.1}$$

$$\lambda_E(t) = \lambda_U(t) \cdot \mathrm{RR}(t) = \lambda_U(t) \cdot [1 + \mathrm{ERR}(t)]. \tag{3.2}$$

If EAR is constant, then exposure to carcinogens increases the incidence of the disease by a constant value for all time periods. When EAR = 0, this indicates no association between exposure to unfavorable factors and the disease.

If RR > 1, then the risk of exposure increases and, conversely, if RR < 1, then it decreases. If RR = 1, there is no association between exposure and disease.

Exposure to carcinogens does not give any new medical or biological effect that would not exist without the influence. It simply increases the probability of effect. So the question is about the level of increase. Exposure to carcinogens can add an additional cancer risk – this is an absolute risk model, that is, an EAR model (3.1). Or it may multiply an existing probability; then this is a relative risk model, that is, an ERR model (3.2).

To make the calculation of the quantitative characteristics convenient, a so-called four-cell table is used (Table 3.1).

The Table 3.1 shows some sample of the population, where a, b, c, and d are numbers of people in the appropriate cell.

Absolute risk (AR) of being ill for persons exposed to the risk factor can be estimated by the ratio $AR_1 = a/(a+b)$. Similarly, for those not exposed to the risk, the likelihood of falling ill is assessed by the ratio $AR_0 = c/(c+d)$.

Table 3.1 Four-Cell Table

	Got Sick	Stayed Healthy
Exposed to the risk factor (the case)	a	b
Not exposed to the risk factor (control)	c	d

The risk can be measured in fractions, in percentage terms or per certain number of people, for example, per 1000 or 100,000. The risk may also take into account the duration of exposure to risk factors and be calculated for a certain period of time of exposure. In other cases, the risk may consider the exposure dose and be calculated per unit of dose.

To compare two groups exposed and not exposed to the risk factor, the relative risk (RR) is calculated. The value of the relative risk allows us to determine the presence and the level of the risk effect on sickness incidence. Relative risk is the ratio of the risk of getting sick in the group exposed to the risk factor to the risk of getting sick in the group not exposed to this factor

$$\mathrm{RR} = AR_1 / AR_0 = [a(a+b)]/[c/(c+d)] = [a(c+d)]/[c(a+b)]. \tag{3.3}$$

Relative risk has a major drawback. Its value depends strongly on the size ratio of the "case" and "control" groups and therefore its use in these studies is meaningless.

It turns out to be more convenient and more reliable to compare not the risks but the parameter named odds.

Odds are defined as the likelihood that the event will occur, divided by the likelihood that it will not happen. In our notation, the probability that there are people among patients who were exposed to the risk factor is $a/(a + c)$. Accordingly, the probability that there are people among patients not exposed to the risk factor is $c/(a + c)$.

The odds of a patient to be the one exposed to risk factors is calculated as the probability ratio $[a/(a + c)]/[c/(a + c)] = a/c$, in accordance with the definition given above.

In the control group of healthy people, the odds of detecting those exposed to a risk factor are $b/(b + d)$, and, for those not subjected to a risk factor they are $d/(b + d)$. Thus, the odds to find someone exposed to the risk factor in the control group are calculated by the ratio $[b/(b + d)]/[d/(b + d)] = b/d$.

Now, the odds ratio (OR) can be calculated by dividing the resulting expressions:

$$\mathrm{OR} = (a/c)/(b/d) = ad/bc. \tag{3.4}$$

OR differs from the ratio of risk. Next, we shall see that for sufficiently rare diseases, the OR is very close to the risk ratio. OR is the only measure that can characterize the disease risk in studies of a case–control type. In cohort studies, risk ratios and ORs can equally be used to assess the hazard of a destructive factor effect.

If $a/c > b/d$, then the factor increases the risk of disease – it really is a risk factor in the common sense of the term. If $a/c < b/d$, then the factor reduces the risk of disease; it is a so-called protective factor.

Table 3.2 Assessment of Lung Cancer Risk in Smokers

		Effect		
		Developed lung cancer	**Stayed healthy**	**Total**
Risk factor	Smokers (Case)	$a = 1500$	$b = 8500$	$a + b = 10{,}000$
	Nonsmokers (Control)	$c = 100$	$d = 9900$	$c + d = 10{,}000$
	Total	$a + c = 1600$	$b + d = 18{,}400$	20,000

ORs can vary from 0 (for $a = 0$) to infinity (for $c = 0$). If the OR is very small (in the limit, OR = 0), then the disease mostly (or only) occurs in the control group. If the OR is close to 1, it means that the disease is virtually unaffected by the risk factor. Finally, if there is a large OR (OR in the limit $\to \infty$), then the disease exists predominantly in the experimental group.

To illustrate these relations, let us consider an example. Let's look at the problem of lung cancer in smokers. Table 3.2 shows some close-to-the-reality numbers of lung cancers among people smoking for at least 20 years.

The first line shows that in a group of 10,000 people smoking, 1500 were diagnosed with lung cancer. The second line shows that in a cohort of 10,000 nonsmoking people, lung cancer showed in only 100 people.

The first line shows that the risk of getting lung cancer after smoking for 20 years is $a/(a + b) = 1500/10{,}000 = 0.15$ or 15%. From the second line, it is easy to see that even if a person does not smoke, the risk of getting lung cancer is $c/(c + d) = 100/10{,}000 = 0.01$, or 1%. The relative risk of cancer from smoking is the ratio of these two values, that is, RR = $a(c + d)/c(a + b) = 0.15/0.01 = 15$. This means that the risk of lung cancer for smokers is 15 times greater than for nonsmokers.

Now let us calculate the OR. The probability that patients who smoke have lung cancer is $a/(a + c) = 1500/1600 = 0.937$. Accordingly, the probability of patients to be nonsmokers is $c/(a + c) = 100/1600 = 0.063$.

In the control group of healthy people, the incidence of smokers is $b/(b + d) = 8500/18{,}400 = 0.462$, and nonsmokers is $d/(b + d) = 9900/18{,}400 = 0.538$.

The incidence of smokers among patients is $[a/(a + c)]/[c/(a + c)] = 0.937/0.063 = 14.9$. On the other hand, the chance of finding a smoker in the healthy group is $[b/(b + d)]/[d/(b + d)] = b/d$. Hence the OR is $(a/c)/(b/d) = ad/bc = 14.9/0.859 = 17.34$. This means that the chance of finding a smoker among patients is more than 17 times greater than the probability of finding a smoker in the healthy control group.

It is seen that the risk ratio is 15.0, and the OR 17.4.

One can calculate the relative difference between the hazard ratios and ORs and determine the sizes for the sample for which this difference is insignificant.

To determine the reliability of the results, the OR parameter can be treated with the entire arsenal of mathematical statistics. Here, we will focus only on the accuracy and reliability of the results, since they often appear in published data on the effect of considered factors on health.

The measure of accuracy is the confidence interval, which can include a calculated value of ORs, whereas an indicator of reliability is the confidence level. Usually, selected confidence intervals equate to a certain number of standard deviations, σ. Problems considered in this book are usually satisfied by the 95% confidence level, which corresponds to a confidence interval of 2σ (more precisely 1.96σ).

However, conventional and well-known techniques for calculating standard deviations are not applicable to ORs, because the distribution of ORs is not normal. Moreover, this distribution is distinctly asymmetrical. Indeed, if a factor is a risk factor, the OR may be in the range between 1 and ∞. However, if it turns out to be a protective factor, that is, it has a positive effect, the OR may vary from 1 to 0. The distribution gets symmetrical with logarithms. Thus, in the case of a negative factor action, the logarithm of the ORs will vary from 0 to ∞, and in the case of a positive factor, from zero to $-\infty$. In mathematical statistics, it is shown that the variance of the OR logarithm equals

$$V=(1/a)+(1/b)+(1/c)+(1/d)$$

Standard deviation is the root of the variance $\sigma=V^{1/2}$ and then the confidence interval for a confidence level of 95% for the log OR is

$$\ln(\mathrm{OR})-2\sigma\div\ln(\mathrm{OR})+2\sigma.$$

Then the boundaries of the confidence interval for the OR can be found by taking the antilogarithm of these values

$$\exp[\ln(\mathrm{OR})-2\sigma]\div\exp[\ln(\mathrm{OR})-2\sigma] \quad (3.5)$$

In scientific literature, particularly in official reports from United Nations Scientific Committee on the Effects of Atomic Radiation (UNSCEAR), National Academy of Science's Committee on the Biological Effects of Ionizing Radiation (BEIR), International Commission on Radiological Protection (ICRP), and others, risk values are usually presented in the form of, for example, ERR = 0.63 (95% CI: 0.52–0.74), where CI is the confidence interval.

3.2 Verification of Tests

To verify the efficiency of tests on various substances for toxicity, mutagenicity, carcinogenicity, and other manifestations, in short, for screening tests that are continuously being developed and introduced, certain criteria are required. Such criteria are called predictive criteria. Here, we describe the predictive criteria, for example, for carcinogenicity tests.

It is assumed that for test verification a certain number of chemicals N, whose carcinogenic properties are already known from other reliable tests, is taken. In addition, it is assumed that of the chemicals in the selected set, N_C are carcinogenic and N_{nC} are noncarcinogenic. Obviously, $N_C + N_{nC} = N$.

Selected chemicals are being tested for carcinogenicity by the investigated test, and it is assumed that the test is considered positive if it detects the carcinogenic chemical, and negative otherwise. It may come out that some substances are detected correctly, and some wrongly. As a result, all substance will be divided into four groups. The number of substances in each group will be given two-letter designations for ease of remembering.

a. Carcinogenic chemicals correctly (True) detected by test as carcinogenic, that is, the test gave a positive response, the number is denoted TP (true positive).
b. Noncarcinogenic chemicals erroneously (False) identified as carcinogenic, that is, test gave a positive response, FP (false positive).
c. Noncarcinogenic chemicals correctly (True) identified, or identified by the test as noncarcinogenic, that is, test gave a negative answer, TN (true negative).
d. Carcinogens erroneously (False) identified as noncarcinogenic, that is, test gave a negative answer, FN (false negative).

The values of predictive criteria are defined as follows:

1. Sensitivity

$$\text{Sens} = TP / (TP + FN) = TP / N_{\text{C}} \quad (3.6)$$

2. Specificity

$$\text{Spec} = TN / (TN + FP) = TN / N_{\text{nC}} \quad (3.7)$$

3. Accuracy

$$\text{Acc} = (TP + TN) / (TP + TN + FP + FN) = (TP + TN) / N \quad (3.8)$$

4. Positive predictivity

$$\text{P.pr} = TP / (TP + FP) \quad (3.9)$$

5. Negative predictivity

$$\text{N.pr} = TN / (TN + FN) \quad (3.10)$$

Sometimes *accuracy* also gets called *concordance*.

The meaning of these criteria is obvious from their names.

In some cases, it is useful to calculate a combination of predictive criteria. If we assume that the calculated values of the predictive criteria are independent random values, the total probability is equal to the product of the probabilities. But the product of the five criteria has some drawbacks. First, analysis shows that the product of the five criteria has a very large dispersion (~100% for 100–200 chemicals). Second, this product is highly dependent on the ratio of the number of carcinogens and noncarcinogens in the chemicals set selected for the verification test. The product of three predictive criteria

$$P_3 = \text{Sens} * \text{Spec} * \text{Acc} \quad (3.11)$$

has a much smaller range and is only slightly dependent on the ratio of the numbers of carcinogens and noncarcinogens.

Although the values of TP, TN, FP, and FN are random variables, their fluctuations are limited because their sum TP + FN = N_C and TN + FP = N_{nC} are fixed values. As it is known, in this case the binomial distribution is found to be true. Thus, absolute deviations of predictive criteria σ are as follows:

$$\sigma(\text{Sens}) = [TP(1 - TP / N_C)]^{1/2} / N_C. \quad (3.12)$$

$$\sigma(\text{Spec}) = [TN(1 - TN / N_{nC})]^{1/2} / N_{nC} \quad (3.13)$$

$$\sigma(\text{Acc}) = \{[TP(1 - TP / N_C)]^{1/2} + [TN(1 - TN / N_{nC})]^{1/2} / N \quad (3.14)$$

Relative deviations of predictive criteria δ are:

$$\delta(\text{Sens}) = [TP(1 - TP / N_C)]^{1/2} / TP \quad (3.15)$$

$$\delta(\text{Spec}) = [TN(1 - TN / N_{nC})]^{1/2} / TN \quad (3.16)$$

$$\delta(\text{Acc}) = \{[TP(1 - TP / N_C)]^{1/2} + [TN(1 - TN / N_{nC})]^{1/2}\} / (TP + TN) \quad (3.17)$$

Relative δ and absolute σ deviation of the product of three predictive criteria:

$$\delta(P_3) = \delta(\text{Sens}) + \delta(\text{Spec}) + \delta(\text{Acc}) \quad (3.18)$$

$$\sigma(P_3) = \delta(P_3) * P_3 \quad (3.19)$$

Real deviation values depend on the number of substances. If the number of carcinogens and noncarcinogens equals 50–100, the average deviation values of sensitivity, specificity, and accuracy equal to about 10%, and ~40%–50% for the positive and negative predictability.

The relative deviation average value for the product P_3 is ~10%.

For the evaluation of the quality of predictive criteria, the χ^2 statistic can be applied. The suggestion to use χ^2 statistics to evaluate the predictive performance of short-term tests was made by G. Klopman and H. Rosenkranz [15] and was used by G. Bakale et al. [16].

3.3 Probit Analysis

The most important and universal characteristic of the impact of various factors on the body is the dose–effect ratio. The concept of dose depends on an impact type. In case of ionizing radiation, dose is consistent with its definition (see Section 1.8.1). Whereas radiation may be short- or long-term, and continuous or fragmented, when exposed to a chemical, the dose may be defined as amount of the substance – usually per unit of mass

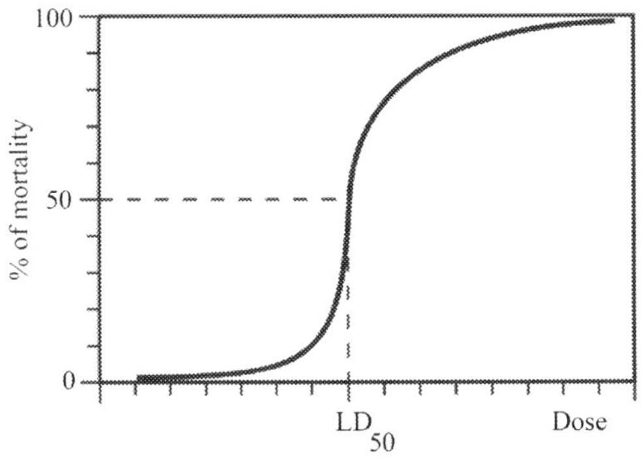

FIGURE 3.1 Typical S-shape dose–response curve.

of the living body – administered orally, or as concentration of the substance in water or air. Also, as in the case of radiation, dose can be distributed in different ways over time.

While studying exposure to radiation or chemicals, different phenomena can be observed, which are called the effect. This may be the death of the organism, or alternatively, survival. It could also be the onset of tumors, induction of various diseases, increased body temperature, a concentration of leukocytes in a blood test, etc.

Basically, the kind of "dose–effect" may differ significantly in different variants of exposure. Dose–effect may or may not have a threshold, and some portions may be linear, sublinear, or superlinear. This dependence can be multimodal, with two maxima, or change in sign as some doses have a positive effect on the body while others, negative. Different types of dose–effect and their features are discussed in detail in the relevant sections. Nevertheless, in many cases, such dependence has a universal appearance, a graph of which resembles the Latin letter *S* as shown in Figure 3.1.

This versatility is quite understandable. First, at a low dose, the impact on the body grows and so does the effect, and then this growth should slow down and finally stop completely. Indeed, the body temperature may not exceed ~42°C, the concentration of white blood cells cannot be more than 100%, as one cannot kill more experimental animals than were taken for research. The curve of "dose–effect" should reach saturation.

Analysis of "dose–effect" allows identifying many important features of the impact of destructive factor on the body.

On the curve, there are several characteristic points. Some points can be distinguished quite reliably; accurate identification of others is far less justified.

The first characteristic point, the definition of which has a greater statistical reliability, is the median value, that is, the point corresponding to the middle of the curve. If the dose–response curve is a mortality curve, the median point is referred to as DL_{50}, which means 50% lethal dose or CL_{50}, 50% lethal concentration. At this dose (concentration), approximately half the animals die in the group. If dose–effect indicates the link between the cancer induction and the dose, then this point is called TD_{50} (tumorigenic dose), which means a 50% chance of tumor formation. Incidentally, the same TD_{50} denotes the median toxic dose, that is, the dose at which the toxicity of a chemical, or its damaging agent, reveals in 50% of cases.

Obviously, the higher the median value, the less dangerous is the damaging agent (chemical or radiation), and vice versa. Thus, sometimes on the x-axis not the dose but its reciprocal is marked. In particular, the toxicity of a chemical is defined as the reciprocal of the half-lethal dose (concentration) $1/DL_{50}$ ($1/CL_{50}$). We also note that the effect usually depends on the duration of exposure to damaging factors. If there are doubts as regards to which measurement option should be selected, the duration gets indicated as the designation of the dose, for example, $LD_{50/30}$ means the median lethal dose for a 30-day exposure to a damaging factor.

Besides the above-mentioned median value, it is desirable to be able to determine other characteristic points. It would be good to determine the absolute lethal dose (LD_{100}), that is, the lowest dose that causes the death of all exposed organisms (ordinate, 100% mortality). It would be useful to know the minimum lethal dose, that is, the highest dose that does not cause the death of any of the exposed organisms (ordinate, 0% deaths). However, since the dose–effect curve both at zero and at 100% is asymptotical and the corresponding analytical values, as will be seen below, are equal to $-\infty$ and $+\infty$, one defines similar values, that is, a dose causing death in 90%, LD_{90}, or 99% of the objects, LD_{99}, and, respectively, the dose causing death of 1%, LD_1, or 10% of objects, LD_{10}, depending which accuracy and reliability of the result is required and can be provided by measurements.

Thresholds are often referred to as Lim. For example, $LD_{10} = Lim_{ac}$ (threshold of acute effect), Lim_{ch} (chronic exposure threshold).

Sometimes other parameters are used, such as intolerable dose, that is, dose less than the deadly but one that causes a fundamental breach of capacity.

Another method for determining the danger of the destructive factor is the angle of the straightened dose–effect curve to the *x*-axis (on the method of straightening, see below). The steeper the line of mortality, the more dangerous is the factor.

One of the most important characteristics of the impact of destructive factor on the body is the maximum permissible concentration (MPC) or the maximum permissible dose (MPD). The legality of the use of such parameters is not evident. Possibly, for some factors, no dose is allowed. Nevertheless, this parameter is widely used in practice, so it is impossible to ignore it. MPD values are determined from the dose–effect threshold Lim_{ch}, and introducing a safety factor K, $MPD = Lim_{ch}/K$. A value for the safety factor of the order of 3–7.5 is selected.

If there is an analytic expression for the function of the S-shaped curve in Figure 3.1, the computational power of modern computers make it easy to calculate any required parameter of dose–effect curve with an accuracy limited only by experimental data. However, we are all prisoners of historical tradition. Analysis of the dose–effect began in the pre-computer era, when calculations were performed on paper, or at least using a slide rule, about which modern young students are probably unaware. To determine the parameters of the curve in Figure 3.1, especially those at the extremities, was quite difficult and not very reliable. This led, as in many other cases, to a linearization of complex curves.

There are several solutions to this problem. The most commonly used, the so-called probit analysis, is based on a normal distribution. The term *probit*, comes from a combination of the words probability and unit – a unit of probability. This method was suggested in 1934 by Chester Bliss in his article dedicated to a quantitative analysis of various poisons' lethal effect, taking as an example the effect of nicotine on oxalic aphids. Currently, this method is widely used in toxicology, radiology, pharmacology, and economics. Bliss proceeded from the fact that the experimental values of the number of dead objects, depending on the dose, is random and normally distributed. It is assumed that there is a binary response of the body: there is an effect or no effect, dead or alive, or sick or well. Then the line effect is an inverted bell curve of individual sensitivities' normal distribution.

Linearization of a dose-response S-curve occurs if the argument is not the dose but the logarithm of the dose lg*D*, and the function is the normal equivalent deviation (absolute deviation in proportions of the standard deviation) $(x-M)/\sigma$ that is, quintiles of the normal distribution corresponding to the effect percentage, or in some other way, the number of sigmas (σ-standard deviation) within which lies the relevant part of the normal distribution integral curve or ratio of the normal differential curve area. Since absolute deviations can have negative values, Bliss suggested to use a special function for analysis, called Probit. Various authors give it different notations, and we will denote it as *Pr*. Probit differs from the normal equivalent deviation by a constant positive number that is greater than the absolute values of the possible deviations, so that the number on the vertical axis always stays positive. Bliss chose for this number 5. So $\mathrm{Pr} = [(x-M)/\sigma] + 5$.

Let us explain this with an example. Value of 16% corresponds to $Pr = 4$, then $Pr - 5 = -1$. This means that in the range from $-\infty$ to $M - \sigma$, 16% of the area under the curve can be placed (or rather 15.87%). The value of 50% corresponds to $Pr = 5$; then $Pr - 5 = 0$.

Typical probit values can be found in tables. You can also determine confidence intervals for the calculated dose values (concentrations).

The logarithm LD_{50} is found from experimental points by the least squares method (or maximum likelihood method) and then, using antilogarithms, one finds the actual value and dose (concentration) LD_{50}. The same method can help to determine other parameters of the dose–effect curve. It is only necessary to understand that the statistical weight of points depends on the degree of remoteness of the median value. Thus, the experimental point corresponding LD_{16} ($Pr = 4.0$), has a statistical weight 1.45 times smaller than the median data point, and the point corresponding to the LD_5 has a statistical weight 3.6 times smaller and therefore a much greater confidence interval. The statistical weight of the experimental points also depends on the number of experimental animals used for individual points.

On the Internet, one can find programs for calculations using probit analysis [17–19].

References

[1] Fletcher RW, Fletcher SW. Clinical epidemiology. The essentials. 4th ed. Philadelphia, PA: Lippincott Williams & Wilkins; 2005.

[2] Biostatistics. A Methodology for the Health Sciences. Ed. by van Belle G, Fisher LD, Heagerty PJ, Lumley T, 2nd edition, 2004.

[3] Background for Epidemiologic Methods. Ch. 5 in: Health Risks from Exposure to Low Levels of Ionizing Radiation: BEIR VII Phase 2, 2006, 132-140 - http://www.nap.edu/openbook.php?record_id=11340&page=132.

[4] Introductory Biostatistics - http://www.medstatistica.com/lib01/stat06.php.

[5] Medical Statistics at a Glance - http://www.medstatistica.com/lib01/stat05.php.

[6] Silva IS. Cancer Epidemiology: Principles and Methods. WHO, International Agency for Research on Cancer, Lyon, France, 1999, 442 p. - http://www.iarc.fr/en/publications/pdfs-online/epi/cancerepi/CancerEpi-0.pdf.

[7] Kellerer M, Nekolla EA, Walsh L. On the conversion of solid cancer excess relative risk into lifetime attributable risk. Rad Environ Biophy 2001;40:249–57.

[8] Environmental Health and Toxicology. IUPAC Glossary of Terms Used in Toxicology. - http://sis.nlm.nih.gov/enviro/iupacglossary/glossarya.html.

[9] Sackett DL, Rosenberg WM, Gray JA, Haynes RB, Richardson WS. Evidence based medicine: what it is and what it isn't. B. Med. J. 1996;312(7023):71–2.

[10] Greenhalgh T. How to Read a Paper: The Basics of Evidence-Based Medicine. 4th ed. New York, NY: John Wiley & Sons; 2010. p. 1.

[11] Slovic P, Fischhoff B, Lichtenstein S. Rating the risks. Environment 1979;2:14–20.

[12] Pigeon N, Kasperson R, Slovic P P. The Social Amplification of Risk. Cambridge, MA: Cambridge University Press; 2003.

[13] Grimes DA, Schulz KF. Cohort studies: marching towards outcomes. Lancet 2002;359:341–5. https://research.chm.msu.edu/Resources/4%20cohort%20studies.pdf.

[14] BEIR VII Phase 2. (2006) p 145. Health Risks from Exposure to Low Levels of Ionizing Radiation - http://www.nap.edu/openbook.php?isbn=030909156X.

[15] Klopman G, Rosenkranz H. Quantification of the predictivity of some short-term assays for carcinogenicity in rodents. Mut. Res. 1991;253:237–40.

[16] Ennever FK, Bakale G. Response of the KE test to NCI/NTP - screened chemicals, III: Complementary value of KE in screening carcinogenesis. Carcinogenesis 1992;13(1992):2059–65.

[17] StatsDirect - http://www.statsdirect.com.

[18] LdP Line - http://www.ehabsoft.com.

[19] PriProbit - http://bru.usgmrl.ksu.edu/throne/probit.

4

Effect of Ionizing Radiation on Biological Structures

Any impact of ionizing radiation on biological objects is usually divided into stages.

1. Physical
2. Physicochemical
3. Chemical
4. Biological

4.1 Physical Stage

A charged particle flying through a substance ionizes or excites its molecules. During ionization, electrons can be transferred energy higher than just ionization energy. The resulting electrons, having excess kinetic energy, are called delta-electrons. Delta-electrons can have sufficient energy to, in turn, create ionization or excitation of molecules. The electrons that during the ionization process either didn't receive enough energy for ionization or have already lost their excess energy by performing ionization are called subexcitation delta-electrons.

The probability of delta-electrons emerging is inversely proportional to the square of their energy:

$$n_\delta(E_\delta) \sim (E_\delta)^{-2}. \tag{4.1}$$

The maximum energy of delta-electrons $E_\delta^{\max}$ for impacting heavy particles with a mass M and kinetic energy E (i.e., with $M >> m$, where m is the mass of an electron) is equal to

$$E_\delta^{\max} \approx (4m / M)E. \tag{4.2}$$

If the impacting particle is a proton, then assuming for simplicity that $M_p/m \approx 2000$, we find that $E_\delta^{\max} \approx E_p / 500$; for an alpha particle, $E_\delta^{\max} \approx E_\alpha / 500$. If the impacting particle is an electron with energy E_e, then an exact calculation gives $E_\delta^{\max} \approx E_e$. Obviously, the resulting ratio is meaningless, because if both the impacting and the knocked particles are electrons, it is essentially impossible to identify which of them is the primary and which is the delta-electron. Therefore, the convention is to consider the one with less energy as the delta-electron. With this approach, the effective boundary of the spectrum of delta-electrons would be the value $E_e/2$.

For analysis conducted in this book, it is essential that the energy of delta-electrons can divert the particle away from its track and produce ionization and excitation at a distance.

Table 4.1 The Ionizing Energy of Water Molecules and DNA Bases in Aqueous Solution [1]

Molecule	Water	Uracil	Thymine	Cytosine	Adenine	Guanine
Ionization energy, eV	12.6	5.27	5.05	4.91	4.81	4.42

The maximum transverse displacement of the delta-electron from the track axis is sometimes used to define the track's radius.

Passing through a biological object, the ionizing radiation interacts with any molecules in its way. The charged particles will mainly touch water molecules, as they occur most often. Major targets are DNA components. Numerical values of ionization parameters are shown in Table 4.1.

In principle, the physical stage lasts as long as the particle moves through the substance until it stops. Typical braking period of particles with energies of the order of MeV, in substances with a density equal to that of water or biological tissue, is several picoseconds ($\sim\times10^{-12}$ seconds) [2]. Many processes, traditionally related to the physical stage of radiation impact on biological structures, end much faster and are usually of the order of 10^{-15} seconds. This means that the wave of physical stage processes moves over a substance with the speed of the particle. While at the end of the track the physical stage is still developing, at the beginning of the track the next stage reactions occur.

By the end of the physical stage, irradiated water contains molecular ions of water H_2O^+, free sub-excitation electrons e^-, excited H_2O^*, and super-excited H_2O^{**} molecules. The list of the products of radiolysis by the end of the physical stage is as follows: H_2O^+, e^-, H_2O^*, H_2O^{**}.

Since in subsequent stages, radiolysis products will undergo transformation, as a result of which their number will change, the final calculation of radiolysis products will be carried out toward the end of the third stage. Obviously, the system at this point is energetically unbalanced.

A charged particle interacts with any cell molecules in its way. Water, in this case, is a model example. This is, firstly, because it is the easiest example of the basic processes of the radiation effect. Secondly, water makes up the majority of a cell, and particle interaction with water molecules is the most likely.

4.1.1 Track Structure

It is essential to understand that, during ionizing radiation, all processes (at least in the first two and, partly, in the third stage) occur not evenly over a substance but in a limited area of space within the particle's influence, called the particle track.

Three factors make the track structure very complicated:

- the statistical nature of the interaction of charged particles with molecules,
- the presence of delta-electrons, and
- the multiple scattering.

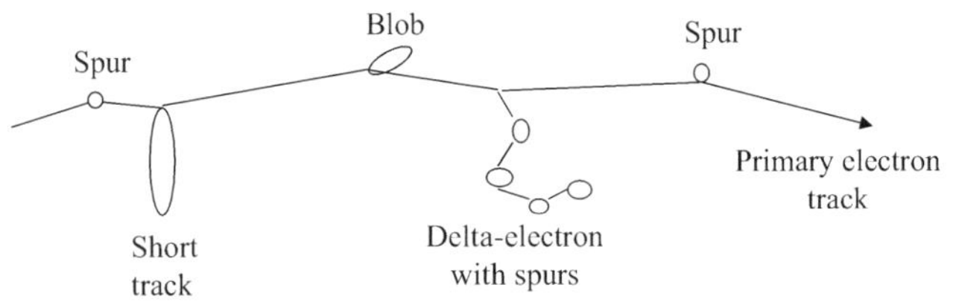

FIGURE 4.1 Schematic track structure of a weakly ionizing particle (fast electron).

Figure 4.1 shows a track of a weakly ionizing particle, that is, a relativistic particle with energy corresponding to the minimum specific energy losses (see Section 1.7.1). Acts of excitation are not marked in the picture. Both collisions with molecules and the energy transmitted in the collision, that is, energy of delta-electrons, are random. Thus, collision points are distributed statistically. In collisions, a particle forms isolated electron–ion pairs and relatively slow electrons that still have enough energy to ionize molecules. The trajectory of slow electrons caused by multiple scattering is significantly curved, and the ionization density is much higher than in the primary particle's track. Obviously, ionization is distributed very unevenly along the track and is concentrated in isolated spurs.

In addition to the spurs, the track structure includes blobs (cells with a large number of ion pairs corresponding to local energy emission of 100–500 eV) and short tracks (cylindrical areas of overlapping spurs with a high density of ionization, formed by delta-electrons with energies in the range 500–5000 eV). The detailed information about track structure can be found in [3–6].

The proportion of spurs, blobs, and short tracks depends on the energy of the primary particle. This dependence is shown in Figure 4.2, based on data from [6].

Ionization cells are distributed in space statistically. The average distance between the cells is determined by the ionization density (LET), which in turn depends on the density of the substance and the charge and energy of the impacting particles.

A track of a strongly ionizing particle is shown schematically in Figure 4.3. It consists of a cylindrical area along the axis at which the bulk of positive ions and excited molecules

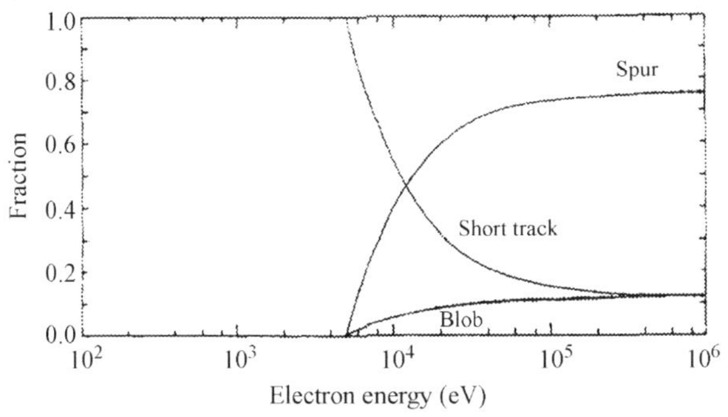

FIGURE 4.2 Energy distribution between spurs, blobs, and short tracks in water in dependence on the energy of the primary particle (fast electron). The figure is from [6] with permission of Am. Chem. Soc. © 1990.

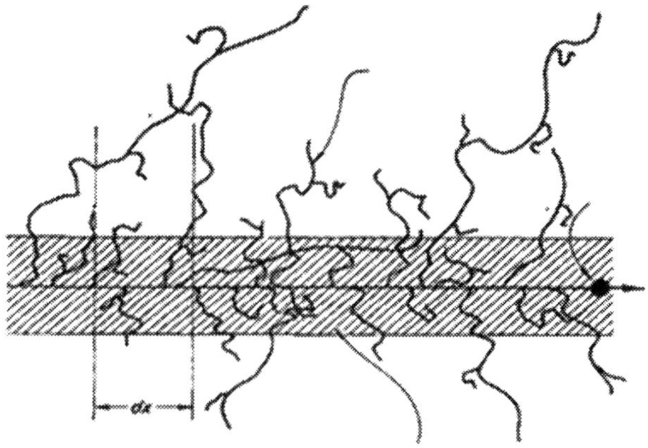

FIGURE 4.3 Schematic track structure of a strongly ionizing particle. The shaded part is the "core" surrounded by the "coat" of delta-electrons.

is concentrated, and on the periphery, the electrons. As well as the central part with high charge density (a so-called core), the track structure also contains the outer part of the track – a "coat" formed by the delta-electrons coming out of the core.

The track radius, that is, the transversal dimension of the area in which the processes of ionization and excitation of molecules occur, can be defined as the maximum transversal displacement of a delta-electron. If a delta-electron effective path is designated as R_δ, then the transversal displacement equals

$$r = R_\delta \sin\theta, \tag{4.3}$$

as shown in Figure 4.4. Here θ is the angle at which the primary particle knocks out a delta-electron. According to classical dynamics, the energy received by the delta-electron is

$$E_\delta = E_\delta^{\max} \cos^2\theta, \tag{4.4}$$

where $E_\delta^{\max}$ is defined by the expression (4.2). That is, it depends on the energy of the primary particle. The electron's effective path is related to its energy according to the relation (1.8)

$$R_\delta = aE_\delta^b. \tag{4.5}$$

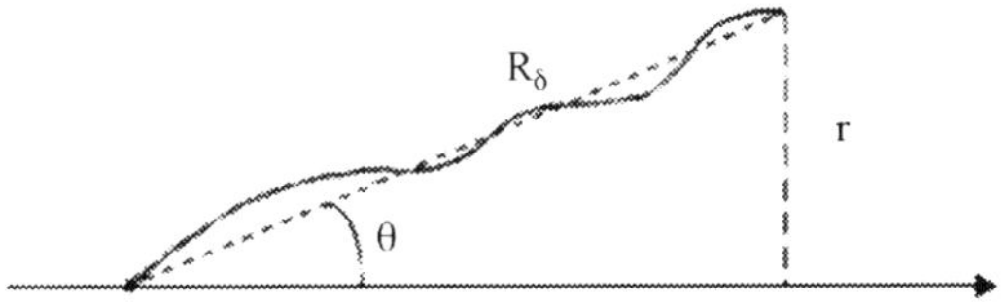

FIGURE 4.4 Definition of the track radius. The direct line is the path of the primary particle; the wavy line is the path of the delta-electron.

Therefore,

$$r = a(E_{\delta}^{\max})^{b}(\cos\theta)^{2b}\sin\theta. \tag{4.6}$$

Strictly speaking, the value b for electrons depends on the energy. For the energy range relevant for the analysis of delta-electron path, a value of $b \sim 1.5$ can be assumed.

With increase of the angle θ, the cosine falls and sine grows. Analysis of the expression for the maximum (4.6) shows that $r = r_{\max}$ at $\theta = \arccos[2b/(1+2b)]^{1/2}$. Hence $\theta = 30°$. From (4.6), it is evident that the radius of the track depends on the primary particle energy and it decreases as the particle decelerates. For an alpha-particle with energy of 5 MeV, $E_{\delta}^{\max} = 2.7\ \text{keV}$. Then, the radius of the track in the initial part of alpha-particle path is about 0.15 μm.

In condensed substances, the ionization density in the tracks of multicharged ions, and particularly fission fragments, is so great that the cylindrical cloud of electrons and ions can be considered as plasma (Debye screening length is less than the radius of the ionization column).

The statistical nature of the distribution of energy in the track and the associated fluctuations significantly influence the effects of radiation on the cell.

4.2 Physicochemical Stage

This next stage of converting the energy of ionizing radiation in a substance is called physicochemical. During this stage, some of the physical processes continue and some reactions between the products of radiolysis start. It lasts for about 10^{-15}–10^{-12} seconds.

4.2.1 Thermalization of Electrons

Let us begin with thermalization of electrons. An electron produced by ionization has excess kinetic energy. Such electrons are called hot electrons. Moving through a substance, a hot electron loses energy in interactions with the molecules of the medium until its energy reaches the level of thermal energy. This process is called thermalization. The thermalization process essentially depends on the type of medium in which it occurs. In atomic gases (and liquids), for example, in gaseous and liquid argon, electrons in collisions with atoms of the medium can only lose energy in elastic collisions. Calculations suggest $\sim 2 \times 10^{5}$ collisions.

In the process of thermalization, an electron traces a complex broken trajectory shaped like a Brownian line and can, in principle, shift by a considerable distance. During electron movement, the parent ion, having received virtually no excess energy during the ionization process, participating in the thermal motion, will move from the place of formation by a negligible distance. Different ionization electrons, having reached thermal energy, will find themselves at different distances from the parent ion. According to calculations, the root-mean-square distance from the point of ionization, to the point where an electron can be considered as thermal one, is about 2×10^{-2} cm in argon at atmospheric pressure. This thermalization process takes approximately 6×10^{-7} seconds.

In molecular substances, a thermalizing electron can lose energy in relatively large portions by exciting vibrations and rotations in molecules. Therefore, the number of collisions needed drastically reduces, the distances passed by an electron during thermalization process get much smaller, and the thermalization process takes less time. This process occurs particularly quickly in polar substances, where a molecule has already a dipole moment, the interaction extends to a greater distance and, hence, the probability of this interaction is higher. In water, the thermalization root-mean-square path is several nanometers [3].

4.2.2 Solvation, Hydration, Self-trapping, Polarons

In any substance, electrons polarize the surrounding atoms and molecules, and thus find themselves inside a polarization field. While moving in a substance (diffusion or drift), the electrons move along with this field, which affects the dynamics of their motion. Most clearly, this effect is manifested in ionic crystals, for example, Na^+Cl^-. Because of the Coulomb interaction, electrons attract positive ions, repel negative, and find themselves in a deep potential well. L. Landau once even suggested that as a result, the electrons must stop. Later, this process received the name "self-trapping." Further measurements showed that in ionic crystals, it's not the electrons that self-trap but the holes. Electrons, being forced to move with the field of polarized ions, become very heavy and slow. Such electrons are called polarons. Their dynamic characteristics are determined by their effective mass, which is noticeably larger than the mass of a free electron. In materials with a smaller share of ionicity of the bond and even in purely covalent materials, the polarization of the charge environment still occurs. Even in liquid helium, because of the strong electron–atom exchange repulsion, a gas bubble forms around electrons. Further, around the positive ion He^+ (or He^{2+}), a sphere with increased density, possibly of solid helium, appears (Atkins's snowball) [7].

The process described here can be characterized, for example, by the depth of the polaron well. For a classic small-radius polaron, the so-called V_K center in alkali halide crystals, it is ~0.2–0.6 eV.

In radiation chemistry, the process of an electron getting into a potential well as a result of the polarization of the surrounding molecules is called solvation, and the electron, solvated. If the environment of the electron is water, electron solvation is associated with the orientation of the polar molecules, and such a process is called hydration and the electron, hydrated. Sometimes, the electron not yet caught in the polarization field is called "dry" and the hydrated one "wet." Hydrated electrons get indicated as e^-_{aq}, and the process of hydration is described as reaction $e^- \rightarrow e^-_{aq}$ [8].

The hydration of electrons formed by radiation in water takes approximately 10^{-12} seconds. Mobility μ and the diffusion coefficient D of hydrated electrons in water is much less than in nonpolar liquids: $\mu(e_{aq}^-) = 2 \times 10^{-3}$ cm²/s·V, $D(e_{aq}^-) = 4.9 \times 10^{-5}$ cm²/second, compared to μ (c-hexane) = 0.24 cm²/s·V, D(c-hexane) = 55×10^{-3} cm²/second; μ (liquid argon) = 480 cm²/s·V, D (liquid argon) ~ 70 cm²/second.

A hydrated electron is characterized by a broad intense absorption band in the red region of the spectrum and a narrow single line (singlet) in the EPR spectrum. The maxima of the optical absorption band of a hydrated electron in water at $T = 298$ K is 720 nm ($E_{max} = 1.73$ eV) [8].

4.2.3 Free Radicals

An important role in the processes inside an irradiated cell is played by so-called free radicals. Free radicals are relatively stable molecules (or atoms) containing one or two unpaired electrons in the outer electron shell. If a radical is charged, it is called the ion-radical.

Radicals are formed as a result of:

- dissociation of excited molecules, such as water:

$$H_2O^* \rightarrow H\bullet + OH\bullet;$$

- During dissociative electron capture, such as a molecule of carbon tetrachloride:

$$CCl_4 + e^- \rightarrow CCl_3\bullet + Cl^-;$$

- In some radiation-chemical reactions, such as water radiolysis:

$$H_2O \rightarrow H^+\bullet \text{ and } OH^-\bullet.$$

Radicals are considered highly reactive particles, because of an unpaired electron. This means the following. If a radical is not charged, then its interaction with other molecules is determined by the same potential as for any neutral molecules. The probability of meeting (or in other words, the collision cross section) for the radical is the same as that for a neutral molecule. But, having met with a molecule, the radical can react immediately at the first collision. The activation energies of such reactions are close to zero.

To illustrate the fact that during nonradical interaction the reactivity may be small, we consider the following example. To characterize electron capture, the so-called attachment coefficient h is often used, which is the ratio of the capture cross section to the collision cross section:

$$h = \sigma_{capt} / \sigma \tag{4.7}$$

The reciprocal of the coefficient h shows the number of electron–molecule collisions required for capture. For example, for a molecular oxygen in the gas phase $1/h = 10^3$–10^4. This means that 10^3–10^4 collisions are required for electron capture. In principle, such a delay in the reaction may be associated with the presence of an energy barrier, or with the searching the ways to return the energy that must be released in the reaction. However, a radical reacts immediately.

4.2.4 Electron Attachment

It is well known that charges with the similar sign of charge repel and with opposite sign of charge, charges attract. Less well known is the fact that there is an attraction between a neutral body and a charge. This is due to the fact that the charge polarizes the neutral body. It induces a dipole moment, which interacts with the charge and leads to attractive forces. So, there is attraction between an electron and any neutral molecule. It is easy to show that the potential energy of interaction between the electron and the neutral molecule depends on the distance between them according to the rule

$$U(r) = -ae^2 / r^4, \tag{4.8}$$

where a is molecule polarizability.

As it is known, in systems with potential interaction energy inversely proportional to the first power of the distance (Coulomb potential), an infinite number of stable states is possible. In systems with a potential of the form $1/r^4$, the number of permitted levels is limited and small, and may even be zero.

Therefore, despite the fact that the attraction takes place between the electron and any molecule, stable negative ions form not with all molecules but only those that lack a small number of electrons to construct a spherically symmetric, entirely filled shell of noble gases.

Substances whose molecules are able to capture electrons to form negative ions are called electronegative or electrophiles, and the ability to capture electrons is known as electrophilicity. The binding energy of the electron in the negative ion is called the energy of the electron affinity U_e.

Table 4.2 gives the values of the electron affinity of some electronegative molecules.

As can be seen, negative ions are very stable formations (i.e., their binding energy is much larger than a kT).

When capturing an electron, affinity energy should be released. The problem of releasing the binding energy is one of the main problems that determine the probability of any reaction of an attachment type. It is a problem we have to deal with when analyzing the recombination process, the formation of molecular ions, and excited molecules, etc. This problem is discussed here with the example of an electron attachment.

One possible way of energy release during attachment is radiation. Consider a thermal electron with energy of the order of kT moving past an electronegative molecule. Radiative attachment should release energy $h\nu = W_e + kT$. The life span before the allowed dipole transition is 10^{-8} seconds, and the time during which a thermal electron, moving at a speed of $\sim 10^7$ cm/second, passes a molecule with a size 10^{-8} cm, is 10^{-15} seconds. Thus, the probability of releasing the binding energy during radiation is of the order of 10^{-7}, which is extremely low. Radiative attachment is an unlikely process occurring in an environment where other processes are even less likely, for example, in the rarefied upper levels of atmosphere.

Table 4.2 Electron Affinity

Ion	H	Cl	O	O_2	C	Guanine	Adenine	Cytosine	Thymine	Uracil	OH
U_e, eV	0.75	3.7	1.5	0.44	1.27	1.51	0.95	0.56	0.79	0.80	1.8

In gases at normal pressures, one of the most probable mechanisms of electron attachment is attachment at triple collisions. Another possible process is a dissociative attachment. For example, electron capture by an oxygen molecule may proceed by the reaction

$$O_2 + e \rightarrow O^- + O.$$

This reaction of dissociative attachment is a threshold one. The electron must have energy of not less than 4 eV, but for some other molecules, dissociative attachment can occur at thermal energy of the electrons. This happens, for example, during the electron attachment to molecules of haloids. The energy of an electron's affinity to a haloid atom markedly exceeds the dissociation energy of the molecule, so that the dissociative capture is energetically favorable at zero electron energy.

Dissociative capture of slow electrons is characteristic for molecules of halogen hydrides (HI), carbon tetrachloride (CCl_4), sulphur hexafluoride (SF_6) and other molecules.

In particular, the neutral water molecule can capture an electron forming a negative ion by the reaction

$$H_2O + e^- \rightarrow H_2O^-$$

or by the reaction of dissociative capture

$$2e^- + H_2O \rightarrow H_2 + 2OH^-.$$

Rate of reduction of electron concentration, n, in a dissociative capture reaction, can be written as

$$dn/dt = -\sigma_{capt} v n N_{im}, \tag{4.9}$$

where N_{im} is the concentration of the impurity capturing the electron, σ_{capt} the capture cross section, and v the velocity of the electrons. Since typically $N_{im} >> n$, then we can assume that the concentration of impurity atoms practically does not decrease, that is, $N_{im} = \text{const}$. Then

$$n = n_0 \exp(-\sigma_{capt} v N_{im}) = n_0 \exp(-t/\tau) \tag{4.10}$$

where τ is the lifetime of electrons before capture:

$$\tau = 1/\sigma_{capt} v N_{im} \tag{4.11}$$

For electron capture to occur, the electron and the molecule must meet, and their motion in the fluid is diffusive. Therefore, the probability of capture depends strongly on the diffusion coefficient. In water, it is two orders of magnitude smaller than in nonpolar liquids. Movement, and the ability to react, for electrons distributed in the solution and, therefore, the reactivity of the electrons depends on the ionic strength of the solution and the pH value.

Velocity coefficients of hydrated electron capture in aqueous solutions were investigated in great details, and for example, the database of Buxton G.V. et al. [9] has the corresponding values for 1515 chemicals.

Electrons in a cell can be captured not only by water molecules but also by other components of living cells, such as DNA bases. The electron affinity energy for DNA bases is shown in Table 4.2. Since such a type of trapping causes, as a rule, the molecule's dissociation, it can lead to a bond rupture in the DNA molecule, that is, to a mutation.

4.2.5 Recombination

Rutherford has already pointed out the fundamental fact that an electron detached from an atom in the ionization process finds itself in the field of the parent ion. Around the ion, there is an area called the capture area or sphere of recombination. The capture sphere radius is the critical distance at which the potential interaction energy of the parent ion and electron equals the thermal energy.

$$e^2/(4\pi\varepsilon\varepsilon_0)r_3 = kT. \tag{4.12}$$

Hence the radius of the capture sphere, sometimes called the Onsager radius, equals

$$r_3 = e^2(4\pi\varepsilon\varepsilon_0)kT. \tag{4.13}$$

For water ($\varepsilon = 80$) at room temperature ($kT = 0.025$ eV), the radius of the capture sphere is 7×10^{-8} cm.

If an electron has gone beyond the sphere of capture during the thermalization process, most likely it is lost for recombination. Thermal motion will throw it away from the sphere of capture. If the electron, during the thermalization process, is inside the sphere of capture, then sooner or later a recombination should occur. Hence it is clear that the possibility of recombination in the electron–ion pair is determined by the electron thermalization process, that is, by how far the electron moved away from the parent ion in the thermalization process. In water, thermalizing electrons will most probably be found in a state of thermal energy within the sphere of capture. Seemingly, recombination is inevitable and the yield of electrons (and hence of the water molecule ions) must be very small. However, it is not quite so.

Recombination gets prevented by two circumstances. First, the electron quickly hydrates and becomes slow-moving. It shifts slowly within the electric field of the parent ion and can join a radiation-chemical reaction before recombination. Second, the electron can be captured by some electronegative molecule inside the cell. In this case, the immersed negative ion also becomes slow-moving in the electric field of the parent ion, and this also reduces the probability of recombination.

The yield of radiolysis products, of course, depends on time, because product concentration changes due to interaction. Values of real output of water radiolysis products are discussed in Section 4.3.2.

The recombination process that determines its kinetics when the electron recombines with the parent ion (as described) is called the preferential recombination (or geminal). If electrons and ions of several pairs are mixed in a spur, the process of recombination occurs in the spur. In the tracks of strongly ionizing particles, such as alpha particles, electrons are in the field of the ion column. In this case, a so-called columnar recombination occurs.

In many complex organic materials, molecular excitation with a high probability leads to nonradiative transitions, and in particular, to their dissociation. This property is widely used in nuclear radiation detectors in Geiger counters and various detectors with a proportional mode of discharge. An additive of organic substances called quenching additive gets introduced into the volume of such detectors as an impurity. Because of quenching gas molecule dissociation, a Geiger counter has a limited life. When all the introduced molecules dissociate, the work of the counter gets disturbed.

In the initial stages of radiolysis, that is, during the physicochemical stage, in most cases the conditions of preferred recombination are fulfilled. After this, the conditions of bulk recombination begin. The mechanisms of bulk recombination are discussed in Section 4.3, dedicated to the chemical stage.

4.2.6 Temperature Rise in the Track Area

Some of the energy of charged particles is converted into heat and this leads to an instantaneous increase in temperature in the track area. In principle, an increase in temperature can be very significant. Thus, the energy released on some of the delta-electron tracks is enough to form micro-bubbles in a liquid. This is the basis for operation of the bubble chamber, a track instrument that played a very important role in the history of nuclear physics. In a pre-overheated liquid, such seed bubbles grow to visible sizes and enable registering traces of charged particles. Bubble chambers worked on liquid propane, freon, hydrogen, and some other liquids.

In the 1950s, to explain a decrease in luminescence yield of organic substances when excited by charged particles with large LET, M. D. Galanin suggested the idea of temperature quenching by increasing the temperature in the excitation channel [10]. Calculations made by Galanin showed that in a channel with a radius of 3×10^{-3}, an alpha particle with energy of 5.3 MeV in anthracene increases the temperature by not less than 100 K.

According to calculations of A. Mozumder [3], particles with small LET (fast electrons) increase the temperature on the track axis in water by about 30 K, alpha particles by about 400 K, and division fragments by 10^4 K. The time during which the temperature on the axis of the track decreases by half due to thermal conduction is 6×10^{-12} seconds for particles with low LET and $\sim 10^{-11}$ seconds for an alpha particle track.

4.2.7 Luminescence

It should be noted that one way of removing excitation in biological structures can be light emission of photons. Human tissue contains large numbers of diverse natural fluorophores that have different spectral ranges of fluorescence. Major endogenous fluorophores in biological tissues include the following groups of substances: flavins, proteins, and porphyrins. Each fluorophore has a typical absorption and emission spectrum. Such proteins as collagen, elastin, as well as intracellular tryptophan, tyrosine, and phenylalanyl, luminesce in the ultraviolet region of the spectrum. Quite a lot of substances fluoresce in the visible part of the spectrum – blue and yellow. These substances primarily include such important components of energy metabolism systems as reduced pyridine nucleotides and oxidized flavoproteins, which occur in any living organism.

Porphyrins are part of hemoglobin, myoglobin, the enzymes catalase and peroxidase, and a large group of cytochromes. All these proteins are involved in the transport of oxygen and energy supply to cells. An ancestor of a number of porphyrins is porphine, a molecule that contains four pyrrol rings, joined into a common interface system by methine bridges. The class of porphyrins includes, for example, chlorophyll and phthalocyanine.

In proteins, aromatic amino acid components can luminesce. Energy absorbed by a part of the protein molecules may migrate through the circuit and transmit to other parts of the same molecule, where it emits. In particular, there are reasons to believe that a change in luminescence can serve as a criterion for the diagnosis of malignancies.

4.2.8 Water Radiolysis Reactions of the Physicochemical Stage

At the end of the physical stage, the following radiolysis products are found in water:

$$H_2O^+, e^-, H_2O^*, H_2O^{**}.$$

Then they undergo following transformations:

a. dissociation of excited and overexcited molecules

$$H_2O^* \rightarrow H\bullet + OH\bullet$$

$$H_2O^* \rightarrow H_2 + O\bullet$$

$$H_2O^* \rightarrow H_2O^+ + e^-;$$

b. recombination of the "dry" electron with a molecular ion, including forming the excited molecule and subsequent dissociation processes

$$H_2O^+ + e^- \rightarrow H_2O^* \rightarrow \ldots\ldots$$

c. electron thermalization and hydration

$$e^- \rightarrow e^-_{aq};$$

d. ion–molecule reaction forming the hydronium ion

$$H_2O^+ + H_2O \rightarrow H_3O^+ + OH\bullet;$$

e. hydronium ion hydration

$$H_3O^+ \rightarrow (H_3O^+)_{aq}$$

Thus, at the end of the second stage, the range of particles in pure water radiolysis is: hydrated electrons e^-_{aq}; radicals H•, OH•, and O•; hydronium ions $(H_3O^+)_{aq}$; and molecular hydrogen H_2.

$$e^-_{aq}, H\bullet, OH\bullet, O\bullet, (H_3O^+)_{aq}, H_2$$

Distribution of these particles in space is not homogeneous. They are concentrated in micro-regions along the particle track. In the case of gamma irradiation, it is the "spurs." Distribution of particles in the "spur" is also uneven. Radicals O•, H•, and OH•; hydronium ions $(H_3O^+)_{aq}$; and the hydrogen molecule H_2 are located mainly in the center of the "spur," and hydrated electrons in the spherical layer at a distance of about 4 nm from the center [11].

4.3 Chemical Stage

Since all chemical reactions that we discuss occur in a solution, the probability of an interaction, that is, reaction probability, and hence the probability of recombination is determined by the mutual diffusion of the reacting partners. To react, they first of all must meet. The rate of such meetings k can be calculated by the Smoluchowski formula

$$k = 4\pi D_{AB}(r_A + r_B) \tag{4.14}$$

where $D_{AB} = D_A + D_B$ is the relative diffusion coefficient, r_A and r_B are radii of components A and B. If one of the interaction components is an electron, its "dimensions" are determined by the De Broglie wavelength.

If the reaction components are charged, one can assume that the radii of the components are equal to the radius of the sphere of capture, $r_{capt} = (e^2 / kT) / 4\pi\varepsilon\varepsilon_0$, and the diffusion coefficient D is related to the mobility of the charge μ by Einstein's relationship $D = \mu kT/e$. Then, from the Smoluchowski formula (1) follows the Langevin-Debye formula:

$$k = 4\pi e(\mu_+ + \mu_-) / 4\pi\varepsilon\varepsilon_0 \tag{4.15}$$

Thus, the Smoluchowski formula becomes the formula for the coefficient of the charge recombination.

In principle, having met, the reacting partners form a diffusion couple and are surrounded by the solvent molecules. In radiation chemistry, this environment is often called the "cell," not to be confused with a living cell. In this cell, they undergo multiple collisions before reaction occurs. If the reaction occurs at the first encounter, such as during the recombination of radicals, such a reaction is called diffusion controlled.

4.3.1 Continuation of Water Radiolysis Reactions

In the chemical stage, chemical reactions between products formed in the previous stage occur. This leads to formation of atomic and molecular hydrogen, hydrogen peroxide, and hydroxyl ions. These reactions occur primarily in the ionization cells, in the "blobs," "spurs," and short tracks. Simultaneously, there is a diffusion of these (and previously formed) particles from the "spur" into the solution volume, resulting in blurring of "spurs" and alignment of the concentrations of radiolysis products in the volume, that is, in establishment of a homogeneous distribution of products. Radiolysis [12] products can interact with each other and with the molecules of the solution. Establishing a homogeneous distribution takes approximately 10^{-7} seconds [11].

The basic reactions of the chemical stage are as follows:

$$e^-_{aq} + H\bullet + H_2O \rightarrow H_2 + OH^-$$

$$H\bullet + H\bullet \rightarrow H_2$$

$$e^-_{aq} + OH\bullet \rightarrow OH^-$$

$$e^-_{aq} + H_3O^+ \rightarrow H\bullet + H_2O$$

$$OH\bullet + OH\bullet \rightarrow H_2O_2$$

$$OH\bullet + H\bullet \rightarrow H_2O$$

$$H_3O^+ + OH^- \rightarrow 2H_2O$$

New molecules that have not occurred in previous stages are hydroxyl ions OH^- and hydrogen peroxide H_2O_2. Subsequently, under certain conditions, the radicals $HO_2\bullet$ and oxygen molecules O_2 can form.

At high excitation density during the chemical stage, an important role is played by counterreactions, when one of radiolysis products gets destroyed and water molecules reappear [11].

$$H + OH \rightarrow H_2O, \text{or } H_2 + OH \rightarrow H + H_2O$$

If a sufficient concentration of radicals is created in the volume, one of the main reactions of radicals is their recombination. In particular, during radiolysis of water, radical ions $H^+\bullet$ and $OH^-\bullet$ form. As a result of recombination of these radicals, a number of highly reactive combinations such as superoxide ($HO_2\bullet$) and peroxide ($H_2O_2\bullet$) appear, which by interacting with a cell's organic molecules can lead to significant damage.

Note also that some of the products of radiolysis are short-lived (hydrated electrons, hydrogen atoms, hydroxyl radicals, etc.) and others are stable (hydrogen, oxygen, and hydrogen peroxide).

4.3.2 Yield of Radiolysis Products

To quantify the yield of products of radiolysis in radiochemistry and radiobiology, the value G is used. G is the number of products per 100 eV (or per joule) of absorbed energy. Physicists are usually interested in the yield of ions and excited molecules. To estimate these quantities, one uses parameters w_i, the average energy for the ion pair production, or $w\phi$, the average energy for photon production. It is easy to see that between these values there is a simple relationship:

$$G = 100 / w_i. \tag{4.16}$$

G was proposed in 1952 by M. Burton at a conference in Leeds. Currently, the yield is sometimes measured in the SI units M/J (mole per joule) [13]. It is easy to find a relation between different units:

$$G(1/100\ \text{eV}) = 0.104G(\mu\text{M/J}). \tag{4.17}$$

Reaction yield depends on the pH, water temperature, LET radiation, and on the time. Yield of radiolysis products in 1 ps (picosecond) after particle passage is called the initial yield, and in 1 μs (microsecond) after, the escape yield. With continuous exposure to ionizing radiation due to the reverse reactions of water synthesis from radiolysis products, a steady state of dynamic equilibrium sets in. Therefore, even in nuclear reactors, water does not undergo complete decomposition. With continuous exposure to radiation, one can talk about stationary concentrations of radiolysis products. However, to assess the impact of water radiolysis on living cells, the initial yield is important. The initial yield in neutral water for gamma radiation is presented in Table 4.3.

With increase of LET from 3 to ~300 keV/μm, the yields of hydrated electrons, hydrogen atoms, and OH• radicals fall and the yields of molecular hydrogen, hydrogen peroxide, and HO_2 grow. Thus, the yield of hydrated electrons in the indicated range of LET changes falls by a factor of about 10, and the HO_2 output increases by a factor of about 100.

4.3.3 Radiolysis of Aqueous Solutions of DNA

In the radiolysis of aqueous solutions of DNA, two main processes occur: depolymerization and the transformation of DNA bases. During depolymerization the deoxyribose DNA fragment gets destroyed. Depolymerization of DNA appears as a decrease of biopolymer molecular mass and solution viscosity. Simultaneously with the decrease of the molecular mass, the hydrogen bonds are broken. This process causes DNA denaturation.

4.3.4 Radiolysis of Aqueous Solutions of Proteins

Basic processes occurring during irradiation of aqueous solutions of proteins are

- polypeptide bond degradation with formation of low-molecular products,
- occurrence of crosslinking,
- reactions between the radical products of water radiolysis and side branches of amino acid residues of the polypeptide chain,
- inactivation of enzymes, and
- denaturation.

Table 4.3 Initial Yield in Neutral Water for Gamma Radiation [11]

Radiolysis Product	e_{aq}^-	H•	OH•	O	H_2	H_2O_2	H_3OZ^+	OH^-	HO_2•
G, 1/100 eV	3.0	0.6	2.8-2.9	0.0067	0.45	0.75	3.3-3.4	0.5-0.6	0.002

4.3.5 The End of the Chemical Stage

Schematic description of transformations of biologically important substances do not describe the entire complexity of the processes occurring in living cells subjected to irradiation. Indeed, after the exposure of the whole body with a dose of 2.5–4 Gy, about 50% of the exposed individuals die within a month. At the same time, chemical transformations in conventional systems at such doses are negligible. Obviously, apart from the indirect effect of water radiolysis products (water forms 85% of living cells), there is a direct effect of radiation on biopolymers. Even taking account of this effect, the biological effects of radiation cannot be explained only by chemical transformations associated with ionization, excitation, and reactions involving free radicals.

During the chemical stage, the components, accumulated by the end of the previous physicochemical stage, react with each other and with the surrounding molecules in solution. As a result of diffusion and chemical reaction, the track spreads and the concentration distribution gradually levels off. Within about 10^{-7} seconds, a relatively homogeneous distribution of products sets in. Sometime later, recombination effectively stops playing any role.

After completion of the chemical stage in water, there are hydrated electrons, hydrogen atoms and oxygen radicals OH•, molecular hydrogen, hydrogen peroxide, hydroxyl, and hydronium ions.

4.4 Biological Effects of Exposure to Radiation

4.4.1 Radiobiological Paradox and Direct Effect

4.4.1.1 Radiobiological Paradox

Before discussing the biological effects of ionizing radiation on living cells, the fundamental paradox of radiobiology should be discussed. Absorption of a negligible amount of radiation energy causes a grave shock to an organism or even its death. For example, irradiation by a dose of 10 Gy, that is, 10 J/kg, received in a relatively short period of time (hours, days), kills any mammal. At the same time, it is easy to calculate that about 10 J of light energy fall to a human body in 1 minute at noon on a sunny day. (However, sunlight is absorbed only by the surface of skin, but ionizing radiation is absorbed in the volume.) In literature, there are also other comparisons. For example, in biological tissue with a volume of 1 μm^3 containing 10^{10} atoms, a dose of 10 Gy produces only about 1000 ionizations. In other words, the lethal dose of 10 Gy affects a negligible number of molecules in a given volume. This dose absorbed by a person weighing 70 kg is equivalent to the thermal energy of only 170 calories. The maximum heating of the human body will not exceed 0.001 °C (the thermal effect is approximately equivalent to drinking a glass of hot tea).

Attempts to solve the radiobiological paradox led to the target theory.

Most of the twentieth century in radiobiology, physical concepts dominated, according to which a cell is a complex of targets. The German scientist, biophysicist, and philosopher Friedrich Dessauer was one of the first to provide a solution. He applied the principles of quantum mechanics and nuclear physics to the problems of radiobiology. He introduced the concept of discontinuity and quantization of energy absorption and the probabilistic nature of absorption events. Dessauer called these events "spot heating" in certain discrete microscopic volumes. Although ionizing radiations have a low average bulk density, individual charged particles have a very considerable energy compared to the energy of chemical bonds, released in a relatively small volume. Probabilistic nature of the manifestation of the radiobiological effect in the cells was due to the statistical distribution of energy absorption events. Now we understand that these local discrete events are molecule ionization acts. The event described, of local release of high energy, was called the hit principle.

Further development of the principle of direct effect is associated with the works of N. V. Timofeev-Resovsky, K. Zimmer, M. Delbrück, J. Lee, and several other researchers (see, e.g., [14,15]).

The main question of direct radiation effect was the following: how many targets one needs to inactivate in a cell to kill it? The concept of one- and two-impact chromosome damage was proposed by Carl Sachs in 1938. Sachs, his contemporaries, and predecessors worked mainly with plant cells. Later, methods of work with mammalian cells in vitro were developed. This made it possible to move from working with plant cells or bacteria to obtaining clear quantitative results with animal cells. The main method of research was to analyze survival curves. We will get to that analysis later (see Section 4.4.4). Note that reliable results are obtained from the survival curves only at relatively high doses.

Effects of radiation on living organisms can be considered on three levels. First, the effect of radiation on molecules is discussed in Sections 4.1–4.3. Secondly, the effect of radiation on cells will be dealt with now. Finally, the effects of radiation on the entire organism in discussed in Section 4.5.

Biological changes under the influence of ionizing radiation may occur immediately after irradiation, during the first few seconds. Such effects are called primary. But some processes can occur with a large time delay, years or even decades later. Such effects are called long-term effects.

Let us start with the primary processes. These include direct and indirect effects of radiation.

4.4.1.2 Direct Action

Direct action is direct transfer of charged particle energy to the functional molecules of a cell. As a result of the processes described for the first stages of radiation effect on a substance (Sections 4.1–4.3), a molecule can be damaged and cease to perform its function. Molecule damages are mainly the breaking of bonds and dissociation. If a damaged molecule is a protein-enzyme or a cell organelle's molecule, it is certainly sad, but possibly not fatal. In synthesis processes occurring in a cell, new molecules can be produced to replace the damaged ones. Therefore, if there aren't too many such damaged molecules in the cell (i.e., the radiation dose was not very high), the cell may well recover and continue its normal activity.

It is a different story if a DNA molecule were directly damaged. As a result, DNA mutation will arise that alters the program of synthesis of other important cell components. Depending on the gene in which the mutation has occurred, it can so substantially change the program that the cell will cease to function correctly. Usually as a result the mechanism of programmed defect cell removal, called apoptosis (see Section 2.3.1), is turned on. Special enzymes dismantle the complex molecules into simpler compounds that are withdrawn through the membrane into the blood and then into the excretory organs.

In principle, damage can be corrected by the repair mechanism. Only a small number of specific mutations does not prevent the cell from functioning normally and does not get fixed by the repair. Such mutations are stored and transmitted to the next generation of cells during mitosis.

Still, most of the damage is corrected by the reparations mechanism. It is very important to note that a cell is a dynamic system in which the processes of damage and recovery are competing.

So, the main intracellular target of ionizing radiation is DNA.

A single event of DNA ionization can seriously disrupt the cell's functioning or even become fatal for it because it is the DNA that contains information on the working program of the cell. This program shows itself in transcription–translation and replication (see Section 2.2). If a cell does not divide (and there are tissues in a living organism in which cells divide very rarely), then isolated defects caused to the DNA will not reveal. So one can assume that dividing cells are mostly affected by radiation. Indeed, this fact was found at the dawn of radiobiology and was named the Bergonie-Tribondeau rule (see Section 2.4).

Currently, we know that fast upgrading normal tissues – gametal cells, cells of blood and the hematopoietic system, epithelial cells of the gastrointestinal tract and skin – are most sensitive to radiation. Radiation damage of these tissues reveals quickly. Neurons and muscle cells show minimal sensitivity. In slowly renewing tissues, radiation damage develops much later, and sometimes, only after additional pathogenic effects. For example, radiation damage of long tubular bones may only reveal in slow fracture healing.

In particular, J. Bergonie and L. Tribondeau concluded that because tumor cells divide often, they should be more sensitive to radiation and, therefore, they can be killed by irradiation. This laid the foundation for radiotherapy. However, the latter statement is true only in part. In fact, cancer cells located in the middle of a tumor may be poorly supplied with oxygen because of the problems with circulatory system development inside the tumor. Such cells remaining in a hypoxic state are less sensitive. This is due to the special role of oxygen during irradiation, which will be discussed in Section 4.4.2.

Some investigators attribute an important role to damage from radiation to the cell membrane [16].

Proteins can become another important target within a cell. They make up between one half and two thirds of a cell's dry weight. It is proteins that act as enzymes – as catalysts of biochemical reactions. Not only the so-called primary structure of the proteins, that is, sequence and type of amino acids in the polypeptide chain, is important to perform their enzymatic function, but also the complex conformation of the molecule, which is called the secondary and tertiary structure of the protein molecule. As a result of exposure to radiation, destructions of amino acids in the chain, chain rupture and violation of the three-dimensional molecular structure are possible. This leads to loss of the enzymatic ability. However, it is known that damage to enzymes requires irradiation at doses much higher than for occurrence of changes in the cell that lead to its destruction. This means that a cell has targets more sensitive to radiation than enzymes.

In conclusion, note that direct immediate action is responsible for only 10%–20% of radiation damage.

4.4.2 Indirect Action

As a result of research of biological effects of cell and organism exposure to radiation, the understanding accumulated that only direct action of radiation cannot explain the observed effects. On this basis, there appeared a concept of a special mechanism for an indirect effect of radiation. One of the reasons for the introduction of the indirect effect mechanism was comparing the effects from radiation exposure to vital macromolecules in solutions and in the dry state. When irradiating enzyme ribonuclease with ^{60}Co gamma radiation, the cell survival marker – the dose D_{37} – is 42 Mrad in the dry state and 0.42 Mrad in an aqueous solution; that is, they differ by a factor of 100 [17]. This suggests that in the aqueous solution an additional mechanism steps in that increases the efficiency of the irradiation. In this example, only 1% of the ribonuclease molecules get inactivated directly by absorption of radiation energy, whereas indirect impacts are responsible for 99% of the effect.

Thus, in the dissolved state in water, macromolecules are by several orders of magnitude more sensitive to radiation than in the dry state. It is natural to assume that products of water radiolysis contribute significantly to damage.

A similar effect occurs with respect to the irradiation of more simple organic compounds, for example, aqueous solutions of formic acid.

An important role in accepting the concept of so-called indirect action was played by the "Dilution effect" (or as it was later called, the "Dale effect"). The essence of this effect is as follows: in irradiation of aqueous solutions of different molecules (e.g., molecules of simple organic compounds, or enzymes), the absolute number of affected molecules does not depend on their initial concentration in a certain concentration range.

For the first time, experiments were carried out in the 1930s by G. Fricke, using solutions of simple organic compounds and in the 1940s by W. Dale with enzymatic solutions.

It is clear that in the case of direct action of radiation, the number of affected molecules should increase with increasing concentration of solute molecules, because of the higher probability of a quantum of radiation to affect them. At the same time, the percentage of the affected molecules should remain unchanged. The experiment although showed that in a certain range of concentrations, the radiation effect did not depend on the concentration of solute molecules. Typical graphs of dissolution influence on the damage nature are shown in Figure 4.5.

Dependence on concentration of the damaging effect of radiation correlates well with the fact that while irradiation in vivo, a certain level of damage of DNA molecules is observed at doses, exceeding by two to three orders those needed to damage these molecules when irradiated in dilute solutions.

Independence of the part of affected molecules from the solute molecule concentration in the solution is attributed to the fact that at a given concentration, not all solute molecules “receive” active products of radiolysis of water. These products are formed in a certain amount at a given radiation dose, but some get intercepted by cellular metabolites and don’t go to biologically active macromolecules. That is, the concentration of solute molecules is not the limiting parameter, but the amount of active water radiolysis products formed at a given dose.

4.4.2.1 Effects of Free Radicals

Since water is the main component of cell content, first of all let us look at the products of water radiolysis. Sections 4.2 and 4.3 showed that radiolysis of water produces free radicals in large concentrations (see Table 4.3), a hydrogen radical H•, hydroxyl radical OH•, the superoxide H_2O_2•, and the hydronium ion H_3O^+. Peroxide radical HO_2• is formed in low concentrations with gamma irradiation, which dramatically increases with increasing LET. These radicals are strong oxidants. Also, an active role is played by nonradical products: hydrated electron e^-_{aq}, which is highly reactive as a reducing agent, and hydrogen peroxide H_2O_2, which, though not a radical, is a very unstable compound and is a source of radical products.

Entering the compounds with organic substances, radiolysis products cause significant chemical changes in cells and tissues, denaturation of protein and other organic structures, and depolymerization of macromolecules, disrupt the permeability of cell membranes, and cause mutations in the DNA and RNA. It is mainly water radiolysis products that determine the indirect effect of radiation.

Note that radiolysis products also form in cells during the normal metabolic processes. As most of these products include oxygen, they are called reactive oxygen species (ROS). To prevent excessive accumulation and to correct the damages occurring, there is a complex set of protective antioxidant systems in the body. Therefore, the damaging effect of radiolysis products is shown at doses at which antioxidant systems can no longer cope with their work.

4.4.2.2 Oxygen Effect

Presence of oxygen enhances the effectiveness of buildup of radiolysis products with oxidizing properties: hydro-peroxide radical HO_2• and hydrogen peroxide H_2O_2. Also, as a result of reactions with radiolysis products, atomic oxygen appears. Because of this, oxygen enhances the damaging effect of radiation.

Radiobiological studies on irradiated cells of various tissues under the presence of oxygen in them showed that the same effect is achieved with much lower dosages than in the case of irradiation in the absence of oxygen. The correlation of doses in these two cases is

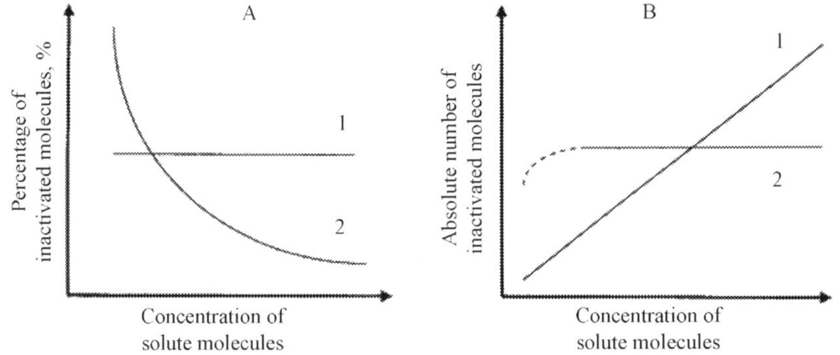

FIGURE 4.5 Typical graphs of dissolution's influence on damage nature. (A) Percentage of inactivated molecules, %; (B) the absolute number of inactivated molecules. For direct (1) and indirect (2) effects of radiation.

called the oxygen enhancement ratio (OER) of radiation damage. The concentration of oxygen in the cell corresponding to its normal content in the atmosphere (~21% = 159 mm Hg) results in an OER of approximately 3 and no longer increases with increase of oxygen concentration up to 100%. With oxygen partial pressure decreasing to 30 mm of mercury, radiosensitivity decreases very slowly and then more sharply. In the range of 3–4 mm Hg, the OER is 2 and then it goes down to 1. It is convenient to show the oxygen effect using as an example the survival curves discussed in Section 4.4.4. Survival curves illustrating the role of oxygen are shown in Figure 4.6, and the connection between OER and the partial pressure of oxygen in Figure 4.7.

Oxygen effect is typical for irradiation of low LET radiation (X-rays and gamma rays). With increasing LET, the oxygen effect decreases rapidly and completely disappears in alpha-radiation exposure with high LET.

4.4.2.3 Other Molecules (Except Water) as Mediators

Originally, in the 1940s, after the emergence of the idea of indirect effects of radiation, only water molecules were considered as molecule-mediators. However, in the 1950–1960s, it was suggested, and confirmed by experimental data, that lipid molecules (primarily those

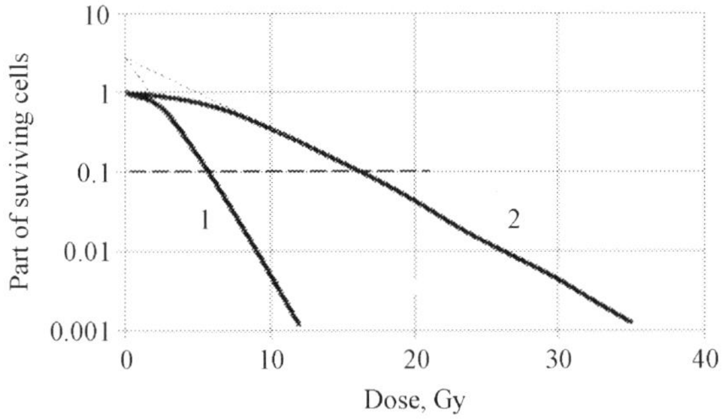

FIGURE 4.6 Typical cell survival curves under X-ray exposure: 1, in air; and 2, in nitrogen; doses causing equal survival rate in nitrogen and air are 15 and 5 Gy, respectively; OER ~ 3.

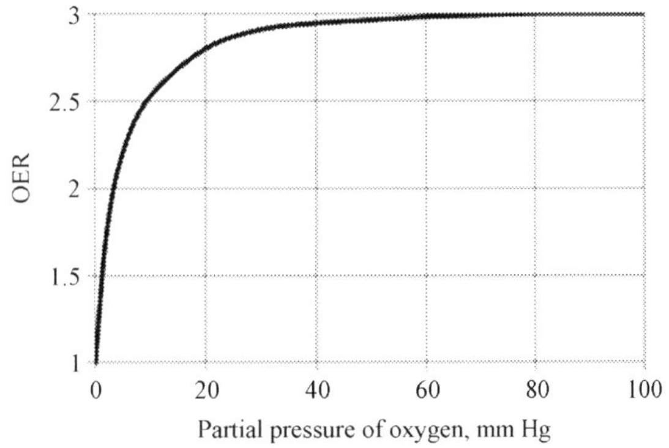

FIGURE 4.7 Dependence of oxygen enhancement ratio on the partial pressure of oxygen at 37°C.

of unsaturated fatty acids) can also act as molecule-mediators. They also form active radical products under the effect of radiation, capable of damaging critical structures and biologically important macromolecules in a cell.

4.4.2.4 Radiotoxins

If the products of cell molecule oxidation by free radical are sufficiently long-lived, they can produce a damaging effect not only on the cell, where they formed, but also, by getting into the blood, they can be transported far away from the place where they were formed and have a pathogenic effect there. Such radiolysis products are called radiotoxins. Lipid radiotoxins, violating the barrier properties of membranes, have the greatest importance in radiation exposure pathogenesis. Radiotoxins formed from certain amino acids may inhibit the activity of many enzymes.

In conclusion, we note that although accurate data on the relative role of direct and indirect effects still cannot be obtained, many researchers believe that the total contribution of indirect effect reaches 90% and is the determining factor in cell inactivation by ionizing radiation.

4.4.3 Bystander Effects

Most of the twentieth century, radiobiology was dominated by the concept that a cell gets damaged only if the charged particle lost its energy in it. Damage could occur as a result of direct action on DNA or indirectly through free radicals, but necessarily in this cell. Damage could be fixed by repair mechanism and then the cell would survive, or this mechanism would fail and then the cells would die. Some of the damage could be identified as mutations that manifest themselves when a cell tries to divide. It was generally accepted that if a cell can pass through five divisions, one can assume that it escaped exposure and its offspring will behave as if it had never been subjected to irradiation.

However, evidence that manifestations of radiation can also be observed in tissues that were outside the radiation field gradually began to accumulate in the middle of the century. Back in 1953, the term "abscopal effect" was introduced by R. J. Mole. The term is composed of the Latin words *ab*, prefix denoting removal, and *scopus*, mark or target. Somewhat later came another term, "bystander effect." Both of these effects mean action of ionizing radiation on shielded tissues outside the irradiation zone, in neighboring or even distant parts of the body. Sometimes, these effects are referred to as "remote." At the moment, it is assumed that the bystander effect corresponds to passing the signal on irradiation from irradiated cells to the nearby non-irradiated ones that were inside the irradiated volume. In the case of abscopal effect, the connection occurs between irradiated and non-irradiated cells that are outside of the irradiated volume. However, some authors use the names of these two effects as synonyms. In English literature, these effects are often referred to as radiation-induced bystander effect (RIBE).

Detailed description of the history of these effects' discovery and acceptance by the scientific community can be found in [18]. The title of the article [18] appearing in the magazine *Mutation Research*, "Changing Paradigm in Radiobiology," is significant. The authors [18] claim that modern radiobiology has undergone a paradigm shift from the hit principle and target theory to the bystander effect.

An idea of the difference of these two effects can be had from Figure 4.8, based on ideas from [19, 20].

Researchers are also finding evidence of information exchange between irradiated cells. The effect of communication between irradiated cells in the irradiated volume is called the cohort effect [19,20].

Remote effects also show in radiation therapy. For example, in a patient with advanced melanoma, local irradiation of one tumor also destroyed another that was outside the radiation field [21].

Existence of remote effects means that the irradiated cell can transfer a signal to non-irradiated cells – bystanders. Apparently, such a signal may be a soluble factor capable of

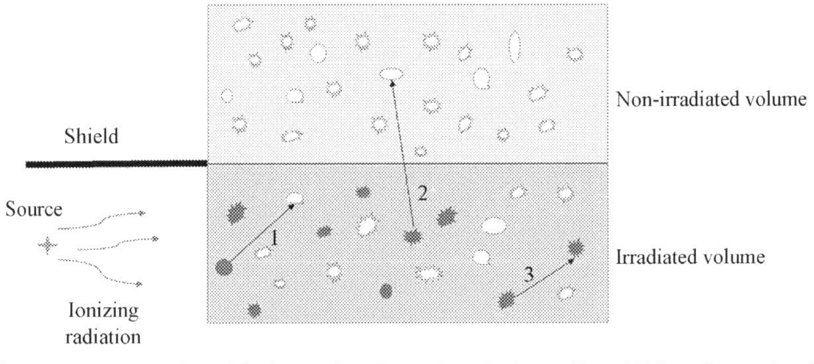

FIGURE 4.8 Schematic representation of the bystander, abscopal, and cohort effects. White cells, non-irradiated; gray cells (red cells in the web version), irradiated. 1, Bystander effect: the interaction between irradiated and non-irradiated cells in irradiated volume. 2, Abscopal effect: the interaction between irradiated cells and cells in nonirradiated volume. 3, Cohort effect: the interaction between irradiated cells in irradiated volume.

moving into the intercellular environment and transmitting information to non-irradiated cells. Possibly, the role of such a factor is played by radiotoxins.

Remote or bystander effects also include the effect of genomic instability (radiation-induced genomic instability [RIGI]). This effect means that no apparent damage can be detected in irradiated cells but they show only in subsequent generations. Genomic instability is observed in vivo and in vitro. However, the detailed mechanism of bystander effects is still not clear.

A detailed description of the contemporary state of research of all bystander effects is contained in the report by the UN Scientific Committee on effects of atomic radiation [22].

4.4.4 Survival Rate

A live cell has two purposes. First, it should perform a certain function in the organ, to which it belongs, and second, it should reproduce the offspring through the division process. Radiation can disrupt both of these functions. Large doses, that is, many acts of ionization and excitation in the same cell, kill the cell and it ceases to perform its function. Smaller doses deprive the cell of its ability to divide. It continues to live and, in principle, can live a considerable time, but loses the ability to divide and eventually dies before mitosis. Even lower doses lead to a delay in the division.

Division processes are organized in time in accordance with a specific cell cycle (see Section 2.2.1). Cell death prior to entering the mitosis phase is called interphase death. If the cell is able to divide, but its descendants are defective and die after one or two divisions, then this is called reproductive death. We have already noted that in animals, cells of some tissues (hematopoietic, genital, and intestinal mucosa) actively divide to produce more of their kind. Cells of other tissues (kidney, liver, heart, muscle, neurons, etc.) divide seldom or never. For all dividing, and most nondividing, cells, an interphase death occurs only at doses of several hundred grays (Gy). Exceptions are lymphocytes and gametal cells at certain stages of their development. They die in an interphase at doses of tens of grays.

As a result of exposure in the cell, one can register a wide variety of reactions. One of the most important ones is delayed division. Numerous studies of various cell cultures showed that on average the division delay is about 1 h per 1 Gy. However, the delay time in each particular case depends on the stage of the life cycle in which the cell got irradiated. The longest delay occurs from irradiation is in the S-phase of DNA synthesis or postsynthetic stage G_2; the shortest is during irradiation in mitosis M, when the vast majority of cells, having started mitosis, finish without delay.

Experiments have shown that the delay of division depends on the radiation dose and is manifested in all cells of the exposed populations, regardless of the subsequent fate of the cells – whether it survives or dies. However, the delay time in different cell types varies. The typical graph of dependence of cell mitosis delay on the irradiation dose for cells in the different phase of a cell cycle is shown in Figure 4.9.

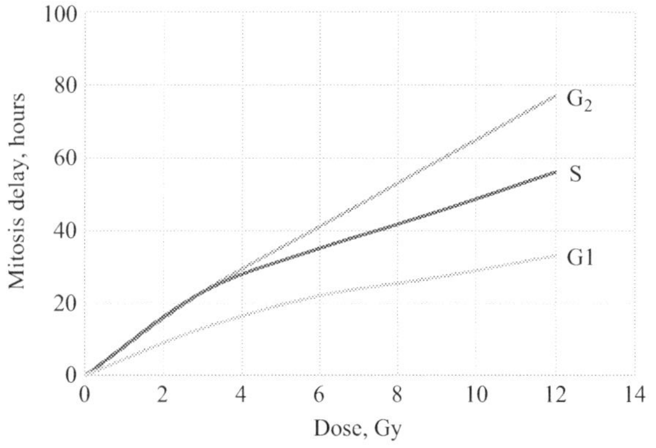

FIGURE 4.9 Dependence of cell mitosis delay on the irradiation dose. Based on [25].

In principle, delay of division can be regarded as a direct effect of exposure, or as a protective response of cells to damage. So far, there is insufficient data to select between these options.

Now, let us analyze cell death in more detail. For many years, the main tools of this analysis were the so-called survival curves. A typical example of dependence of the percentage of surviving cells on the dose is shown in Figure 4.10. In this figure, the y-axis shows the proportion of surviving cells in logarithmic coordinates and the x-axis the dose of gamma-irradiation in linear coordinates. In such coordinates, exponential dependence should be demonstrated by a straight line. The graph shows that at low doses the relationship is curvilinear, and then with increasing doses it becomes linear; that is, it turns into an exponential dependence.

It should be noted that such curves are generally obtained *in vitro* from mammalian cell cultures. The percentage of surviving cells is determined based on the number of normal healthy colonies of cells grown in irradiated plates compared to the total number of colonies in the control plates. Obviously, in this case, only those cells were described as surviving that passed through multiple divisions, that is, preserved their reproductive potential.

Curves under consideration show not the degree of cell damage (the degree of damage is always the same – cell death) but the probability of death. This makes an essential difference between the situation with death of cells and the situation with division delay, where the effect (as opposed to effect probability) depends on the dose.

On a typical survival curve such as in Figure 4.10, one can specify three parameters: shoulder D_q, the number of hits n and survival D_0. At low doses, the curve has a shoulder. Then the intersection of the line parallel to the x-axis and coming out of the top of the curve with the extrapolation of the linear part of the curve gives a value of shoulder D_q. It is believed that this is the measure of the ability of cells to repair. Extrapolation of the linear portion of the curve to the intersection with the ordinate axis gives the extrapolation number n, characterizing the number of targets (hits) that is necessary to destroy the

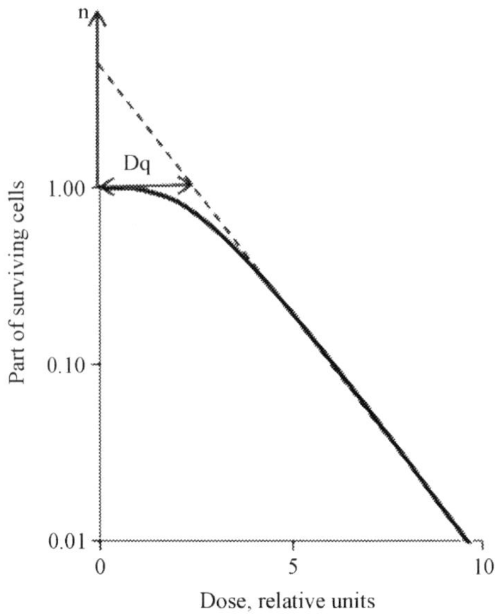

FIGURE 4.10 Typical survival curve. The shoulder D_q and extrapolation of the inclined part of the curve up to the intersection with the *x*-axis are pointed out.

cells. The linear portion corresponds to the exponent and, in principle, can be described by the relation

$$N(D)/N_0 = \exp(-D/D_0). \tag{4.18}$$

Here N_0 and $N(D)$ are the number of cells before and after irradiation, D_0 is the parameter of the curve that defines its slope and is called the survival rate, and $1/D_0$ is the radiosensitivity. Another parameter used to describe the survival curves is D_{37} – the dose at which the percentage of surviving cells decreases by *e* times, so that 37% of them remain. If the survival curve was pure exponential, then D_0 would be equal to D_{37}. But they differ because of the shoulder; the value of D_{37} is greater than that of D_0.

For the majority of dividing cells, D_0 = (1.2–2.0) Gy, D_q ~ 1.5–5 Gy. The radiosensitive fraction of hematopoietic cells in bone marrow has a much greater radiosensitivity D_0 = 0.1 Gy, for the lymphocytes D_0 = 0.5 Gy.

In particular, the survival curves reflect the radiosensitivity of cells (see Section 4.4.6). Significantly, radiation sensitivity depends on the cell cycle phase in which the cell was irradiated. In other words, radiosensitivity varies over time quite noticeably, by a few dozen times. Cells prove to be most radiosensitive during mitosis. Further change in radiosensitivity is somewhat different in different types of cells. However, as a rule, it is minimal (and survival rate is maximal) when irradiated in the late S stage, when the radiosensitivity is approximately 10 times lower than in mitosis.

In principle, the survival curves can also be built in other coordinates, where the ordinate shows not the number of surviving objects but of those that died. Then the well-known

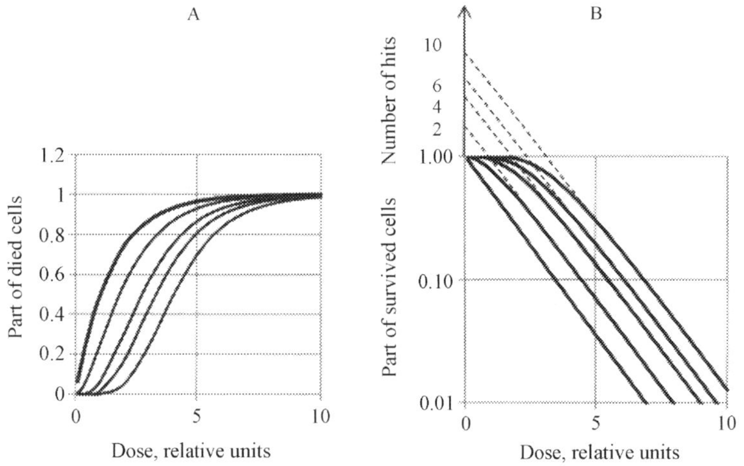

FIGURE 4.11 Dose survival curves for objects with different target hit rates. (A) Death rate in the usual scale (S-shaped or sigmoidal curves); (B) Surviving rate in semi-logarithmic scale (hits number is indicated on the curves).

S-shaped curve appears. Both versions of curves are shown in Figure 4.11. Curves in Figure 4.11A and B correspond to each other. From this figure, it can be clearly seen how the forms of S-shaped curves change with the growth of the shoulder.

The curves with the shoulder can be described on the basis of the so-called multitarget model:

$$N(D)/N_0 = 1-[1-\exp(-D/D_0)]^n. \tag{4.19}$$

Here, n is the number of targets that the particles must hit to inactivate the cell. With $n = 1$, the relation (4.19) becomes (4.18).

During gamma irradiation, one event of interaction of a fast Compton- or photo-electron with a DNA molecule can lead to single chain rupture. Single ruptures themselves usually do not cause cell death. Repair system quickly and effectively corrects this damage. However, two different fast electrons can accidentally produce a double chain rupture, that is, a breakage of both DNA strands. In this case, the cell is, with high probability, doomed. In fact, sometimes one fast electron can produce a double rupture. It is believed that during gamma irradiation, for about every 100 single ruptures, there is 1 double one.

With an appreciable probability, double ruptures occur during passing of a strongly ionizing particle. That's why, high-LET radiation is much more dangerous.

Based on these concepts, cell survival rate in many cases can be described using the so-called linear-quadratic model of Chadwick and Leenhouts. According to this model, cell survival rate is described by the formula

$$N(D)/N_0 = \exp[-(\alpha D+\beta D^2)]. \tag{4.20}$$

Here α and β are parameters characterizing the probability of DNA ruptures in an irradiated cell. In this model, it is believed that single ruptures are proportional to the first

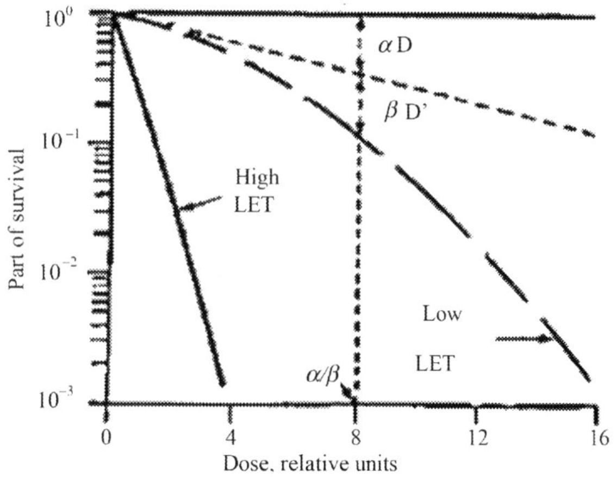

FIGURE 4.12 Survival curve in the case of linear-quadratic (α/β) model.

power of the dose and double ruptures to the square. The corresponding curve is shown in Figure 4.12.

Curves with shoulder are obtained using low LET radiation: gamma- and hard X-ray radiation. With growing LET, the shoulder decreases. In the case of alpha-particle radiation, the curve turns into a proper exponent. In this case, one hit of a cell nucleus by a particle for DNA structure damage is enough.

If exposure to gamma or X-ray radiation occurs with low-dose or after radiation exposure and a few hours' pause is made, to enable a repair of damaged DNA structure, the percentage of surviving cells significantly increases. This is due to the creation of conditions for activation of enzymatic repair systems and increased DNA structure recovery time prior to synthesis before the entry of cells into mitosis. Depending on conditions, the recovery rate is typically of time intervals from 1 to 10 hours.

The role of repair processes is illustrated by the graph in Figure 4.13, which shows the influence of the dose rate on the shape of the survival curves.

To compare different exposure conditions, one uses the parameter "radiation exposure efficiency," which is the relationship between dose LD_{50} (dose for 50% mortality) at short-term exposure and high dose rate to the same value under different conditions of exposure. With decreasing dose rate, below a certain level, the radiation exposure efficiency falls and for irradiation taking several days it can drop several-fold. It is believed that in this case, rapid recovery processes at the cellular and tissue levels begin to show.

Repair of damage and increased percentage of surviving cells also occurs at the so-called fractionated irradiation. A typical pattern of possible survival curves in the case of short-term irradiation sessions separated by pauses lasting from several hours to several days is shown in Figure 4.14.

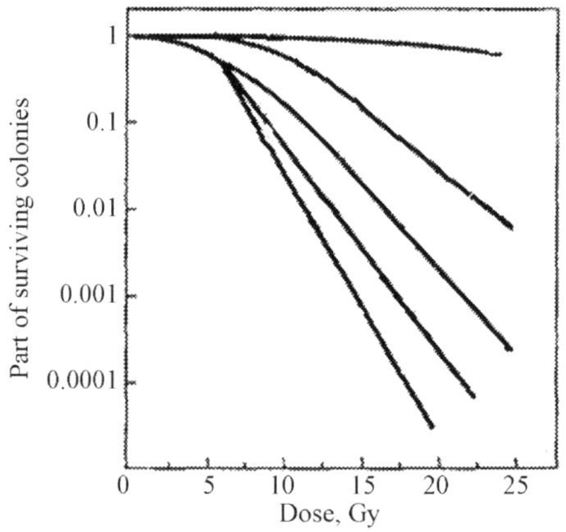

FIGURE 4.13 Survival in dependence on a dose with increasing radiation dose rate downwards from ~0.5 cGy/minute up to ~1 Gy/minute.

4.4.5 The Dependence of Radiation Exposure on LET

In the history of research of the effects of ionizing radiation on biological objects, the most common and studied form of radiation was X-ray radiation with a maximum energy of 250 kV. However, when study of other types of radiation effects on cells and tissues began, it became clear that the same effects on the same cells are obtained at different doses. Particles with higher LET create more damage.

For proper consideration of the impact of various types of radiation, the relative biological effectiveness (RBE) was introduced

$$RBE = \frac{D\ (\text{any effect for X} - \text{ray radiation with energy of } 250\ keV)}{D\ (\text{any effect for another type of radiation})}$$

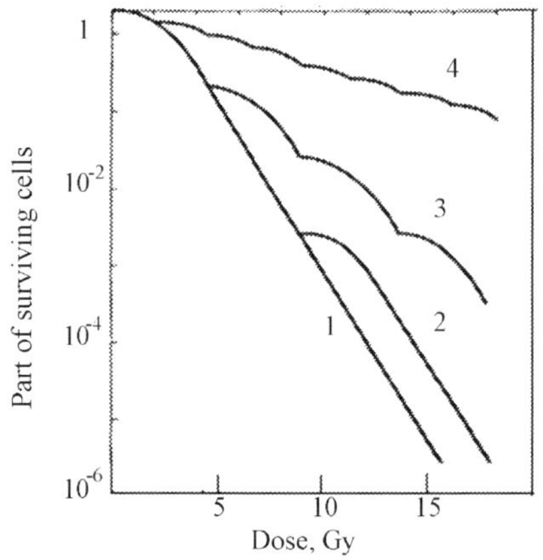

FIGURE 4.14 Survival curves for the fractionated radiation exposure. Curves 1, single exposure; 2, 3, and 4, exposure is divided into two, four, and eight equal fractions, respectively.

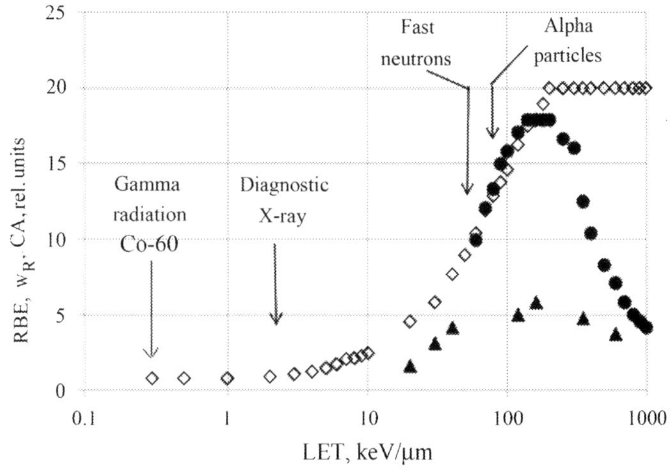

FIGURE 4.15 Connection of radiobiological effectiveness (RBE) of radiation, radiation weighting factors w_R (quality factors), and lens epithelium chromosomal aberrations (CA) with the linear energy transfer (LET). Empty diamonds, QF according to ICRP 26; circles, RBE; triangles, CA. The values for CA are from [23].

Different types of radiation having impact on biological objects mainly differ in their LET value. Therefore, the value of RBE depends on the LET.

The connection between LET and RBE values is shown in Figure 4.15. It is seen that the relationship between these two values is not strictly linear, but a positive correlation of LET and RBE is definitely observed to a certain maximum value of LET, corresponding to ionization density on the tracks of alpha particles. RBE decreases at high LET because of the fact that a particle with such a large local release of energy can kill more cells than it can reach on its track.

The same figure shows the regulated values of quality coefficients for different types of radiation. Also shown are the results of experimental studies on the yield of chromosomal aberrations (CAs) in the lens epithelium. In preparation of this figure, data from [23] was partly used.

In dosimetry, to calculate equivalent doses, the so-called quality factors (or radiation weighting factors) of any given type of radiation are used. Values of quality coefficients are closely related to the RBE, but strictly speaking, they are not equal. It is important to understand that the RBE values strongly depend on the experimental conditions (type of cells, type of radiation, dose rate, etc.). Quality coefficients are fixed values that are used to calculate dose rates under any conditions. Quality coefficients are the operating parameters used to solve practical problems. On the other hand, RBE is a scientific parameter, taking into account measurement conditions.

Officially accepted values of quality coefficients are shown in Table 1.1.

4.4.6 Radiosensitivity of Tissues, Organs, and Organisms

Radiosensitivity is a measure of an organism's reaction to ionizing radiation exposure. The reciprocal of radiosensitivity is radioresistance. Radiosensitivity, as a concept, is very important both from the theoretical and practical points of view. Knowledge of the mechanism controlling radiosensitivity and its regulation will enable conscious control of tissue response to irradiation, weaken it to protect the body, and strengthen it during radiation therapy of malignant tumors or bacteria inactivation.

Different cells, tissues, organs, and organisms react differently to radiation and produce different specific effects. In summary, the existing rule states that cell radiosensitivity is proportional to division frequency and inversely proportional to the cell specialization degree. That is why, frequently dividing and unspecialized cells, as a rule, are the most radiosensitive. Most obvious examples of such cells are marrow and embryo cells. On the other hand, central nervous system (CNS) cells of an adult are highly specific and have very low or zero division frequency. So CNS cells are relatively radioresistant.

An example of extremely low radiation sensitivity is bacteria found in a nuclear reactor channel. Under these conditions, the bacteria not only died but lived and multiplied. That is why they were called *Micrococcus radiourens* – radioresistant micrococcus.

Values $1/D_0$ or LD_{50} (a dose at which 50% of organisms die) may serve as a quantitative measure of radiosensitivity (see Section 4.4.4). Usually, in this case the typical time period of observation is also specified. The typical value of such a period is 30 days after irradiation and the corresponding parameter is denoted $LD_{50/30}$. For a human, this value is of the order of 3.5–4.4 Gy. However, the active use of the achievements of modern medicine, especially bone marrow transplantation, enables pushing this limit toward higher doses. For example, here are several more values of $LD_{50/30}$. For a triton, it is 30 Gy; for a turtle, 15 Gy; for a rat, 7.1 Gy; and for a dog, 2.75 Gy.

A special feature of radiation damage to a living body is that such a body cannot be regarded as a simple collection of cells and organs. Being part of a tissue, cells are very dependent both on each other and on the surrounding environment. A response to the damaging effect of radiation involves all the body's control systems – nervous, endocrine, etc. Different organs have different radiosensitivity, which depends not only on the cells comprising the organ but also on their location, vascularity, and even irradiation time. Introducing the concept of critical organs enables slight simplification of the situation. Critical organs are the vital organs or systems that are the first ones to fail within the studied radiation dose range, resulting in death of the organism within a certain period after irradiation. For more about critical organs in humans, see Section 4.5.

To consider different sensitivities of different organs and tissues in the calculation of effective dose, so-called weight coefficients are introduced. Values of weighting coefficients for different organs and tissues are shown in Table 1.2.

4.4.7 Long-Term Consequences

Long-term consequences of exposure are phenomena that occur years and decades after exposure. They can be of two types: somatic, developing in exposed individuals, and genetic, hereditary diseases developing in the offspring of the exposed parents. The most dangerous and most widespread long-term consequence is cancer. The mechanism of the

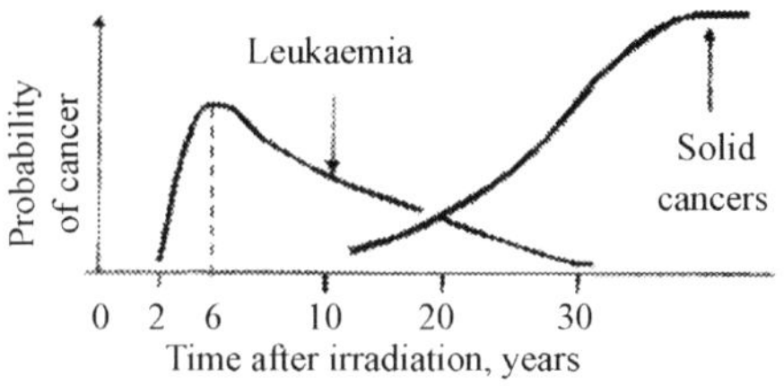

FIGURE 4.16 Probability of cancer depending on time after exposure.

carcinogenic effects of radiation is described in Section 2.4. Dependence of the probability of cancer from the time elapsed after exposure is shown in Figure 4.16.

From Figure 4.16, one can see that the first type of cancer with a delay of only a few years is leukemia. Leukemia shows within approximately 20 years after irradiation, and then almost completely disappears. In particular, leukemia was the first type of cancer found in the survivors of atomic bombings. Radiation is more likely to induce leukemia than all other cancers. So leukemia is often considered the "marker" of radiation effects.

Tumors or, as oncologists call them, solid cancers, have a longer latency period. They rarely appear in the first 10–20 years, but after that, new cancer cases can be diagnosed for at least 50 years.

Besides cancer, there are other possible manifestations of long-term effects of radiation on the human body: physiological disorders (thyroid malfunction, etc.), cardiovascular diseases, allergies, chronic respiratory diseases, immune deficiency and associated increase in sensitivity to infectious agents, and temporary or permanent sterility.

We note, in particular, two types of long-term effects: damage of cell fiber constituting the lens, and as a consequence clouding of the lens (cataract), and reduced life expectancy unassociated with cancer. Experimental studies on mammals suggest that a proportional relationship exists between the degree of shortening life expectancy and radiation dose. Every 0.01 Gy of single irradiation reduces life expectancy by 1–15 days, and at chronic exposure, by 0.08 days.

Yet it should be clear that the likelihood of long-term effects is experimentally proven only at relatively high doses. Reliable detection at low doses is difficult because of problems with gathering enough statistical material and adequate control groups of animals. In addition, and most importantly, it is difficult because of the huge background of similar diseases in humans caused by other environmental factors, including carcinogenic and mutagenic. For more information about the role of small doses, see Chapter 6.

4.5 Radiation Sickness

In the mid-twentieth century, after the Second World War, humanity learnt about a new scary-sounding and dangerous radiation sickness. Various troubles associated with radiation became known immediately after the discovery of X-rays and radioactivity. But the first-in-the-world fatal case of radiation sickness occurred on August 21, 1945, with the

American physicist who worked at Los Alamos, Harry Daghlian Jr. Much better known is the second case: the death from radiation sickness of the Canadian physicist Louis Slotin in a Los Alamos lab working on developing nuclear weapons. The life of Louis Slotin and his death from radiation was reflected in the novel "The Accident" by an American writer Dexter Masters. In this novel, Slotin features under the name Louis Saxl.

At the same time, victims of the nuclear bombings of the Japanese towns of Hiroshima and Nagasaki on August 6 and 9, 1945, became known. It is impossible to identify the proportion of people that died in the explosion just from radiation, because the people who were close to the epicenter got exposed to a blast wave similar to that of a very powerful but nonnuclear bomb, and to strong light radiation. Many died under the rubble and from many other explosion factors. But after the explosion, doctors could already distinguish between a death due to burns or wounds and due to radiation sickness. By December 1945, 177 people died from radiation sickness [24]. Radiation sickness struck hundreds of residents of the Marshall Islands and the crew of a Japanese fishing boat *Fukuru-Maru* as a result of thermonuclear bomb tests in 1954. Humanity was faced with radiation sickness, and because of sharp confrontation between the great powers during the "cold war" and the danger of global thermonuclear slaughter, people feared its further massive appearance. So, an intensive study of this new phenomenon began.

Note that there is a certain convention: total body irradiation at a dose of less than 1 Gy, and locally at larger doses, is called "radiation injury" or sometimes "acute radiation reaction." At doses higher than 1 Gy, radiation sickness develops.

Up to this point, we have learnt about the effect of ionizing radiation on molecules, cells, and tissues. The problem of radiation sickness leads us to the analysis of the effects of radiation on the entire organism. It should be borne in mind that an organism is not simply a sum of cells but a complex dynamic system of interacting cells. Previously (see Section 4.4.3), it was shown that cells are able to pass each other information about irradiation (bystander effect, abscopal effect, genomic instability effect). In particular, cells transmit information to each other about the need to start active division (e.g., for healing wounds) or to stop active division (when the integrity of the body is restored). Activity of cells is run by whole-body hormonal and nervous systems. One of the manifestations of cell operation control is apoptosis – a genetically programmed cell self-destruction in response, in particular, to external signals. To characterize interactions in a cellular society, the term "social behavior of cells" is even used.

So, as a result of irradiation by a dose not lower than a particular level, radiation sickness starts. There are two variations of radiation sickness: acute radiation syndrome (ARS), after a short intense irradiation, and chronic radiation sickness (CRS), after long-term exposure at a relatively low dose rate.

Radiation sickness is a multisymptom disease and its specific development and dominating symptoms strongly depend on how the exposure occurred – what dose rate and how it was distributed in time, that is, simultaneous, fractional, or continuous – over a period of time. If continuously, the dose rate is important. For radiation sickness development, it is important whether irradiation was external or internal, and whether the whole body or only part of it got irradiated.

At doses above a certain threshold, radiation sickness is manifested by symptoms, the severity of which increases with increasing dose. These manifestations are called deterministic. The lower threshold of deterministic effects depends on many conditions as listed above. Each effect and each organ has its own threshold dose. With overall short-term irradiation in a human, the threshold dose for deterministic effects is from 0.2 to 1 Gy.

As is known, dividing cells often are the most radiosensitive (the Bergonie-Tribondeau rule). Therefore, organs with a large number of such cells get affected by irradiation primarily and at lower doses. These are hematopoietic organs, epithelium of intestines and stomach, ovaries and testes.

Organs primarily affected by radiation are called critical. Depending on which organs are mostly affected and which of them are the most critical, there are several forms of radiation sickness: bone, brain, intestinal, and cerebral.

Various forms of radiation sickness can be easily illustrated by a curve showing the dependence of life span on the dose. Such a curve for a single X-ray irradiation of mice, developed by B. Rajewski in 1953 [25], is shown in Figure 4.17.

On the curve, one can clearly see stepwise changes in life expectancy from changing the dose, associated with the failure of critical systems. Similar curves were later obtained for other mammals (rats, hamsters, guinea pigs, monkeys, and humans) as well as for amphibians, insects, worms, and even plants. However, the specific dose rates at which the corresponding critical organ died differed for different animal species, pointing to the different radiosensitivity of the responsible systems. Thus, in this case we are dealing with a general biological law.

The portion of the curve corresponding to low doses (from a few to ~10 Gy) is responsible for the death of animals due to failure of bone marrow (hematopoietic) system. In the dose range of ~10–100 Gy, there is a plateau where the average life expectancy is about 4 days. This Section corresponds to defeat of the epithelium of the small intestine. The survival time for gastrointestinal death is dose-independent. With further increase of the dose, the life expectancy drastically shortens, down to several hours, which is associated with damage to the central nervous system. The fact that some parts of the curve correspond to the specific system destruction is confirmed by conclusive and easily reproducible experiments.

The presence of the plateau in the curve (Figure 4.17) indicates that damage to a system incompatible with life occurs at a certain level of exposure, that is, it has a threshold nature.

Experts divide all the possible manifestations of radiation sickness in humans into periods, phases, variants, and stages in great detail. More detailed description of the radiation sickness can be found, for example, in the manual [26]. We will look into the most common variant of radiation sickness – acute radiation sickness caused by relatively even irradiation relative to the body's volume. Phases of acute radiation sickness are very clearly shown in Figure 4.18, taken from [27].

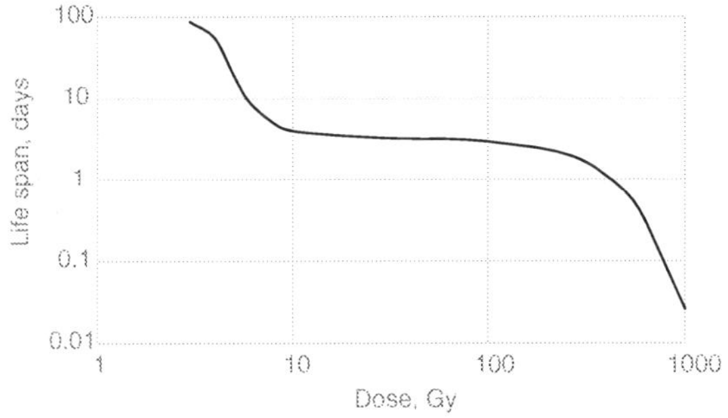

FIGURE 4.17 Dependence of the average life span on the dose for mice. Based on [25].

As shown in Figure 4.18, immediately after irradiation there is a short delay; its duration depends on the dose and can take from several minutes to several hours. After the delay, the primary reaction phase occurs. Then, the painful symptoms subside and even disappear and the phase of apparent clinical well-being (hidden or latent phase) begins. The duration of this phase is also dependent on the dose; it takes from several hours to 3–4 days. Then again, the clinical manifestations return and the manifest illness stage starts. The disease ends with recovery or death.

It is useful to know the indicators of primary reactions:

- nausea and vomiting;
- lack of appetite;
- dryness and bitterness in the mouth;
- a feeling of heaviness in the head, headache, weakness, drowsiness;
- shocklike state;

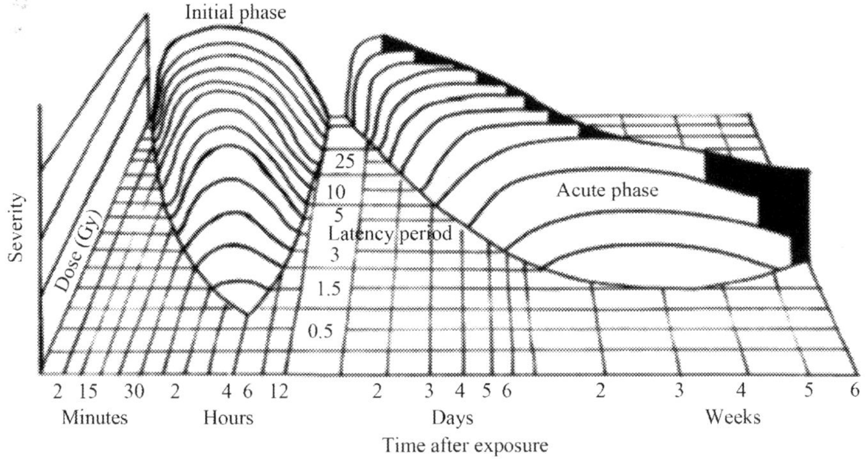

FIGURE 4.18 Stages of acute radiation sickness. The figure is from [27] with kind permission from Editions John Libbey Eurotext, Paris.

- drop in blood pressure;
- brief loss of consciousness;
- diarrhea;
- fever;
- neutrocytosis and lymphopenia in the peripheral blood during the first day after irradiation.

Type of damage is also affected by the method of exposure: the effect of an external ionizing radiation source (external radiation) or radionuclides that got inside the body (internal radiation). With a uniform distribution in the body of incorporated (internal) caesium-137, the disease picture is similar in clinical features, with radiation sickness arising due to external irradiation. Because of the difference in chemical properties of radionuclides, they can become very heterogeneously distributed in the body and cause selective organ failure: strontium-90 accumulates mainly in bones, plutonium-239 in lungs, iodine-131 in the thyroid gland, etc.

It is obvious that with increasing doses, disease manifestations grow: increased temperature, the frequency of the urge to vomit, or the concentration of leukocytes or DNA fractions in the blood along with other indicators, accumulation in the cells of the oxidation products of biological membranes (peroxides and hydroperoxides of unsaturated lipids), etc. At low doses, this growth is almost linear; then it slows and soon stops altogether, the curve reaching saturation. This is the so-called S-shaped or sigmoidal curve corresponding to a quite natural, universal form of dependence. Indeed, at very small doses, the higher the dose, the greater its impact. But there are limits. Body temperature cannot exceed ~42 °C. Concentration of white blood cells is limited. Patient death is the ultimate event.

Such an S-shaped curve describes the mortality of a living body depending on the irradiation dose. It is shown in Figure 4.19. It can be seen that the curve showing the dose/effect relationship does not start at zero but is shifted to the right, toward higher doses. That is, there is a threshold – the minimum lethal dose corresponding to death of the organism.

Values of LD_{50} are different for different species of mammals but the shape of the dose curve and causes of death are the same. At doses around LD_{50}, the hematopoietic system is critical for the body; at high doses, the critical system is the mucosa of the small intestine. In the first case, part of the animals die within 10–14 days; in the second, 4–7 days after irradiation. At $D > 1$ Gy up to the absolutely lethal dose, radiation sickness of various severities is observed in survivors.

At lower doses, no apparent clinical manifestations are observed but there may be long-term effects (see Section 4.4.7), mainly cancer (see Section 2.4), cataract, and disorders of the nervous system and others.

Further, irradiation can affect both somatic and gametal cells; that is, it can cause damage to the genetic apparatus. In both cases above, at a certain dose, deterministic effects can be observed, and in the genetic apparatus, only stochastic. Stochastic effects are analyzed in detail in Section 6.

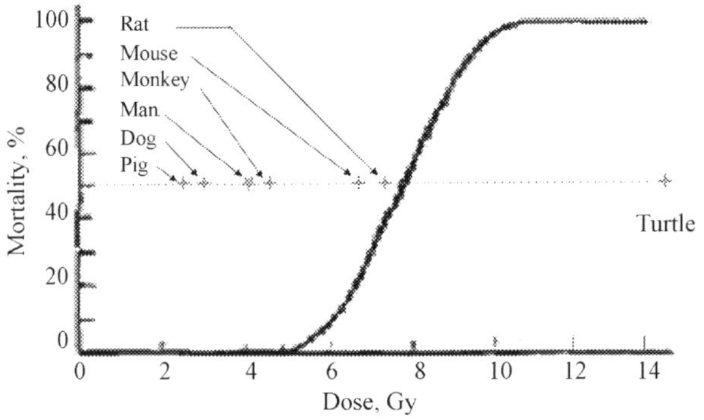

FIGURE 4.19 Dose curve of mammals' death from gamma-irradiation of the whole body. Crosses show the values of LD_{50} for different animals and a human.

Finally, one more type of radiation effects on the body – the so-called teratogenic (from the Greek *teratos* [monster]) effects of irradiation – should be pointed out. Teratogenic effects of radiation are manifested in congenital malformations in children irradiated at the in utero development stage. These are not genetic but somatic effects arising from exposure of the foetus.

4.6 Radon and Internal Exposure

4.6.1 Properties of Radon

We devote a separate section to radon, as it is of particular interest.

First of all, radon is the main source of internal exposure of the body. Other nuclides get into the organism in significant concentrations only in special cases – during accidents at work in contaminated areas, etc. Radon, in greater or lesser concentrations, is always there regardless of our desires and behavior.

Second, radon is a good and long-known example of the positive effect of ionizing radiation. Radon baths are widely used to treat many types of diseases. Spas based on natural sources of radon waters were popular long before the discovery of radioactivity and continue to attract people and really help them.

Radioactive water baths have been applied even before radon discovery in Jáchymov, Czech Republic, and in Bad Gastein, Austria. Eighteen spa companies and health spas in Germany, Austria, and the Czech Republic organized the EURADON–European Association radon spas (Verein Europäische Radonheilbäder e.V.) in 1988 and jointly promote the use of naturally occurring radon as a means of treating medical conditions. Hundreds of people from all over the United States and Canada journey to six "Radon Health Mines" located in the mountains of Montana (USA). On the other hand, the role of radon as the main cause of lung cancer, for example, in uranium miners, is of no doubt. It would seem that it is undoubtedly an example of hormesis. Where is the line between beneficial and harmful effects of radon? We will try to highlight the current state of this problem.

The decay chain of ^{238}U is shown in Figure 1.5. It can be seen that radon occurs during the alpha decay of ^{226}Ra (one of the members of the chain) and in turn undergoes alpha decay. Subsequent products in the chain can undergo both beta and alpha decays. Of these, we distinguish two alpha decays: ^{218}Po and ^{214}Po – the first element with a half-life of 3 min, the second with fractions of a second. The last of them turns into a ^{210}Pb with a half-life of 22.3 years. Then, the decay chain is delayed for an appreciably greater period than the duration of radon decay product influence on human health. In the later part of the decay chain, the famous nuclide ^{210}Po with a period of 138.4 days appears. This has become well known to the public through recent extensive newspaper articles about Litvinenko's poisoning. ^{210}Po in the process of alpha decay is converted into the final nuclide of the chain – the stable lead isotope ^{206}Pb.

Radon is a member of the family of noble gases. This means that because of the properties of the electron shell of the atom, its energy bonds with other radon atoms and any other atoms is so small that thermal motion at room temperature breaks this bond easily. This is evident from the fact that radon liquefaction temperature is –61.9°C. Therefore, inhaled radon does not stay in the lungs and gets almost completely exhaled.

This is different with the alpha-active decay products of radon ^{218}Po and ^{214}Po. In the lungs, they are deposited in the respiratory tract and undergo alpha decay until the lungs clear of the inhaled dust and aerosols naturally.

In external air, new atoms arising from the decay of radon atoms very quickly (<1 second) bind with molecules of gases and vapors contained in the air and form tiny microparticles with a diameter of about 1 nm. These microparticles within 1–100 seconds connect to existing aerosols with a diameter of up to 1 μm. Distribution of aerosols by size, their structure and properties, have a strong influence on their filtering in respiratory ways and thus the dose received by the body from breathing air that contains radon.

Radon has a short-lived isotope, a member of the decay chain of ^{232}Th, called thoron. Its half-life is 55.6 seconds. Because of its short lifetime, thoron is much less likely to emanate into the air from a substance containing thorium compounds. In thorough analysis of doses, the role of thoron is also considered, but it can be neglected in the estimates.

4.6.2 Inflow of Radon Into the Atmosphere

The most important natural source of ^{222}Rn is the decay of ^{226}Ra in soil and rocks in the Earth's crust. The average content of ^{226}Ra in soil is equal to 2×10^{-12} g of radium per 1 g of soil.

According to the secular equation (1.3), in principle, equilibrium between uranium and radium should be established. However, because of the long lifetime of radium, this equilibrium can be broken as a result of the difference in chemical properties, because of some thermal processes etc. Measurements show that, for example, in granite sands, the balance between uranium and radium is considerably impaired, and radium concentrates on the surface of the mineral particles.

As a result of radioactive decay of radium, radon atoms enter into the minerals' crystal lattice and then into pores and cracks. To describe the output of radioactive gases, two terms are used. The process of release of radon from porous solids of relatively small size is called "emanation." The process of release of all gases from vast geological structures, from rocks, soils, etc. is called "exhalation." So, radon exhales from rocks and then emanates into air.

The percentage of radon atoms emanating from a solid substance is determined by the emanation coefficient. Typical coefficient values for emanation from rocks and soils vary in the range from 0.05 to 0.7.

For an atom of radon to inflow from a solid material, the "mother" radium atom during the decay must be close to the surface or to a pore in the rock. Maximum distance at which the recoil energy from the decay can throw atom is 20–70 nm in most typical minerals, 100 nm in water, and 63 μm in air. Radon emanation and transport process is the subject of extensive studies [28,29].

Radon activity is measured in units of activity per time unit, for example, becquerels/year or becquerels/day. The activity of rock materials relates to a unit of rock mass (Bq/t s). Emanation of radon into the air is measured in units of Bq/m^3h.

Most radon that is released from soil and rocks in the earth's crust is transferred to the surface and carried into the environment by groundwater flows. Intensive leaching of radionuclides out of the rock leads to formation of radioactive water in some areas, including water containing radon-222 in quantities greater than $1.85\ 10^5\ Bq/m^3$ (5×10^{-9} Ci/L).

Despite significant differences in radon emanation conditions, some typical values can be indicated as a guideline. Radon flux out of dry soil approximately equals 0.033 Bq/m^2s; in wet soil, it is lower by about a half.

Radon concentration in mineral waters varies very significantly. Thus, in the Barbanstein minefield (Austria), radon concentration reaches 2.2×10^6; in Baden-Baden (Germany), 7.8×10^5; and in Yakhimov (Czech Republic), $6.3 \times 10^6\ Bq/m^3$.

Because radon is a product of uranium decay, its inflow is affected by the concentration of uranium in soil and rocks. In addition, the output of radon depends on rock fracturing and the position of the groundwater level. Output of radon is also affected by the time of the day and season. Therefore, the yield of radon in different locations can vary significantly. As a result, radon activity in different environments can be characterized by the following average figures (Table 4.4).

Radon and its short-lived decay products are the most important sources of natural human exposure. They account for more than half of the average annual dose. Miners have long been

Table 4.4 Radon Activity in Various Sources

Radon Source	Activity, kBq/day
Natural gas	3
Water	4
Outdoor air	10
Building materials and ground under the building	60

aware of the health risk from inhaling radon. The role of radon in the environment started to be taken into account only after 1970, when it became clear that in some cases, the concentration of radon in indoor air can be comparable with the concentration in underground mines.

Since damage in cells is produced by alpha particles and the effect depends on their total energy, the degree of exposure to radon is measured in units of energy of alpha particles (potential alpha energy concentration [PAEC]), in J/m^3. For individuals associated with inhaling radon only at work, such as uranium mines or radon therapy, for historical reasons, a special unit is used: WL (working level). Usually, they talk about the monthly working level (WLM). WLM is a measure of the total radiation dose received by a miner from inhalation of 1 working level within 1 working month (170 hours, that is, approximately 22 working days of 8 hours). Total energy of alpha particles corresponding to 1 WLM equals 1.3×10^{11} eV per liters of air within 170 hours. Hence, it is easy to calculate that 1 WLM is equivalent to 3.54×10^{-3} J h/m^3 or 6.37×10^5 Bq h/m^3 [30].

Monthly working level can be easily calculated if the activity of radionuclide is measured. However, in many cases, activity is not known and then the degree of exposure is determined indirectly from assumptions about the relation of radionuclide concentrations and degree of equilibrium in the decay chain.

Yield of radon into the air depends on the permeability of the building material of the building's foundation, the foundation's pressure on the soil, the type of backfill under the foundation, floor material, and the gap between the floor and walls. It may range from about 0.1 to about 100 Bq/second. In calculations, it is assumed that radon exhalation rate in the soil under a building is 37 kBq/m^3 hour. Then, at typical building design conditions, the rate of exhalation into indoor air is 1–10 Bq/m^3 hour. If a typical exchange rate of indoor air is 1 hour^{-1}, the concentration will be 1–10 Bq/m^3.

Estimates show that radon concentration in indoor air is contributed to by 40% diffusion through building materials, 40% by air convection bringing radon from the basement, and 20% from outside (from the environment).

Another source of radon in a home is water from the public water supply. Radon gas emerges from the water during use. Typical values of radon concentration in water depend on where the water is taken from. Well water has highest concentrations ($\sim$100 kBq/m^3), groundwater has medium concentrations ($\sim$10 kBq/m^3), and surface runoff is minimal ($\sim$ 1 kBq/m^3). The ratio of the concentrations of radon in water and in the air is taken to be 10^{-4}, although in some cases it can be much larger (by a factor of 70).

Radon concentration depends not only on its release but also on meteorological conditions. In rainy weather, it falls markedly. It also depends on the type of housing, ventilation conditions, and lifestyle and behavior of the population.

At night and early in the morning, radon keeps closer to the surface of the earth. In the day, when heated by the sun, radon rises. During the day, the radon concentration can vary by a factor of 10. Seasonal changes are also possible, especially in areas covered in certain seasons by snow and ice. Typical concentration of radon in the atmosphere varies between 1 and 100 Bq/m^3. The UNSCEAR experts take 10 Bq/m^3 as an average value [31]. A small number of thoron concentration measurements give a value of 0.1 Bq/m^3.

Calculations of doses produced by radon and its decay products are strongly dependent on the equilibrium in the decay chain. At equilibrium, the equivalent concentration is calculated by the formula $C_{eq} = 0.105C_1 + 0.515C_2 + 0.380C_3$ where C_1, C_2, and C_3 are the concentrations of decay products ^{218}Po, ^{214}Pb, and ^{214}Bi, taking into account the contribution of each of the decay products into the total energy of alpha particles. With this in mind, the measured concentration of radon can be converted into the equilibrium equivalent concentration (EEC), directly proportional to PAEC.

The relationship between EEC and PAEC is determined by the equation $1\ Bq/m^3 = 5.56 \times 10^{-6}\ mJ/m^3 = 0.27\ mWL$ [30].

Numerous measurements in many countries all over the world give the average value of radon concentrations in rooms in typical conditions of about 30–40 Bq/m^3. To illustrate the effect of particular conditions on the concentration, let us give an example. In the premises of the Brussels museum, which displays the uranium-bearing minerals from Katanga, radon concentration is about 10–15 kBq/m^3, in spite of the increased ventilation. This is several hundred times greater than in normal premises. Because of radon and gamma radiation of minerals in the house of a museum caretaker who lives nearby, the absorbed dose was estimated to be 6.5 mSv/hour (~45–50 mSv/year). Shielding the minerals enabled reduction of this dose by a factor of about 5.

4.6.3 Radon's Effect on Health

The defining feature of measuring doses produced by inhaled air containing radon is that it is impossible to insert a dosimeter into the trachea or, rather, into some critical surface epithelial cells of the respiratory tract. Therefore, the doses have to be calculated by formulating specific models.

The most important variables that must be considered when calculating doses received by the nuclei of cells from alpha particles are the size of aerosol particles, respiratory rate, and depth of the target cells inside the tissue.

Freshly formed aerosol particles have a size in the range of 0.5–5 nm and form an ultrathin or, as they say, nonadherent fraction. Most of these particles are attached to the aerosols with a size of 20–500 nm and form an adherent fraction. The ratio of the adhering fraction depends on aerosol concentration. In dusty and smoky air, almost all the ions produced by the decay of radon are included in the adhering fraction. In clean air, the nonadherent fraction may dominate. Usually in a nonadherent fraction, the ratio of nuclides is $^{214}Pb/^{218}Po = 1/10$.

Respiration rate of an adult male on average is 0.45 m^3/hour in the state of rest (8 hours per day) and 1.2 m^3/hour in an active state (16 hours per day). Similar figures for women are 20% lower for the state of rest and 5% lower for the active state. Recently, these values have been revised, and the new estimates of respiratory rate is 22.2 m^3/day for adult men and 17.7 m^3/day for adult women.

When radon and its decay products get into the lungs, the basal and secretory cells of the upper respiratory tract get exposed to alpha particle radiation. But since alpha particle ranges are very small, it is important to know exactly which cells are the main critical targets.

Average depth of the cells responsible for cancer was determined surgically as 27 and 18 μm for basal cells and mucus, respectively.

Considering the latest and most accurate data of various physical and biological parameters, dosimetrists calculated the absorbed doses from inhaling radon. For some selected typical parameters and reasonable dosimetry models, considering the equilibrium concentration distribution of radon decay products, the effective dose for both indoor and outdoor ranges within 5–25 nSv m^3/Bq hour. The distribution maximum is 9 nSv m^3/Bq hour or 5.7 mSv/WLM.

The annual effective dose from radon indoor is ~1.0 mSv/year and outdoor ~0.095 mSv/year.

Numerous studies in both case-control and cohort options show that high doses of radon cause lung cancer. It was officially recognized by the World Health Organization (WHO) in 1986. In the report by UNSCEAR experts [30], obvious evidences of the connection between the risk of lung cancer and exposure to radon and its derivatives were presented. Most data on the risk of lung cancer induced by radon was obtained in epidemiological studies of miners [32].

Applying data obtained from miners to the risk assessment of lung cancer in the population causes considerable difficulties. One needs to consider many factors when extrapolating from the conditions in the mines to the conditions in homes. Among these factors are uncertainty of the linearity of the dose-effect, the difference in risk for adult men (miners) and the population, including women and children, the difference in other potential effects of environmental pollutants such as arsenic, quartz, engine exhaust gases, and many more. One must take into account the difference in the degree of equilibrium of radon decay products and special aspects of breathing of working miners and of much less intensively working population.

The report of the International Commission on Radiological Protection (ICRP) [32] provides data on studies of the effects of radon on the induction of lung cancer in the population over the period 1990–2006. Results of 20 studies performed in the United States, Europe, and China were presented. The result is expressed in units of relative risk per 100 Bq/m^3. Part of the studies provide values of relative risk of less than 1; the other part of more than 1. However, the excess over 1 is very small, with wide confidence intervals, and in almost all the papers the final limit is below 1. All this means that no clear link between radon and lung cancer in the population was found.

Since the beginning of the 1990s, American scientist B. Cohen has conducted extensive environmental studies of the impact of radon in dwellings on mortality of residents of 1729 U.S. counties from lung cancer. The research was so extensive that it became possible to obtain reliable dose dependence. This is shown in Figure 4.20; the graph has been built using data from [33]. The research covered almost all the white population of the United States (89% of the population, or about 200 million people). Findings of the research show reduction of mortality from lung cancer in U.S. counties with concentration growth in homes [33–36]. Thus, in this case, a radiation hormesis can be observed (see Section 6).

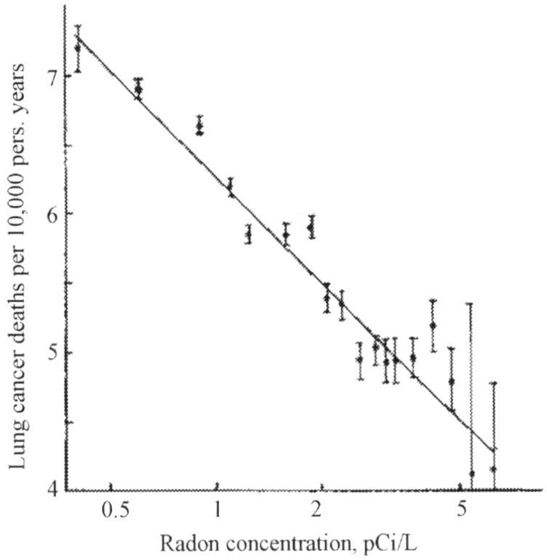

FIGURE 4.20 The dependence of mortality due to lung cancer among U.S. population on radon concentration in indoor air. The data are from [33].

The data of B. Cohen shows that mortality in miners grows with increasing dose at the same doses as those received by the population because of much larger dose rate in mines. A similar increase was observed in the staff of plutonium production at the "Mayak" plant (located near Ozyorsk city in Russia, 150 km south-east of Ekaterinburg and 72 km north-west of Chelyabinsk). However, we must bear in mind that cancer of the respiratory part of lungs is mainly associated with plutonium, and bronchial cancer is associated with radon.

There are reasons to believe that the body's defence against radon is associated with stimulating the formation of corresponding repair enzymes for DNA, damaged not only by radiation but also by other harmful agents widespread in the environment.

References

[1] Crespo-Hernández CE, Arce R, Ishikawa Y, Gorb L, Leszczynski J, Close DM. Ab initio ionization energy thresholds of DNA and RNA Bases in gas phase and in aqueous solution. J. Phys. Chem. A 2004;108(30):6373–7.

[2] I.M. Obodovskiy. Complex of Problems on Experimental Methods of Nuclear Physics. Energoatomizdat, 1987, p. 79, problem 1.22. [in Russian. (И.М. Ободовский. Сборник задач по экспериментальным методам ядерной физики. М.: Энергоатомиздат, 1987, с. 79, задача 1.22)]

[3] Mozumder A. Charge particle tracks and their structure. Adv. Rad. Chem. 1969;1969(1):1–102.

[4] Charged Particle and Photon Interactions with Matter: Chemical, Physiochemical and Biological Consequences with Applications. Ed. by Mozumder A, Hatano Y. CRC Press, Boca Raton, FL, 2003, p. 860.

[5] Charged Particle and Photon Interactions with Matter – Recent Advances, Applications, and Interfaces. Ed. by Hatano Y, Katsumura Y, Mozumder A. CRC Press, Boca Raton, FL, 2011, p. 1045.

[6] Pimblott SM, LaVerne JA, Mozumder A, Green NJB. Structure of electron tracks in water. 1. Distribution of energy deposition events. J. Phys. Chem. 1990;94:488–95.

[7] Khrapak AG, Schmidt WF, Illenberger E. Localized Electrons, Holes and Ions. Ch.7 in Electronic Excitations in Liquified Rare Gases. Eds. Schmidt WF, Illenberger E. ASP, 2005, 239-273.

[8] Hart EJ, Anbar M. The Hydrated Electron. Wiley Interscience 1970;267.

[9] Buxton GV, Greenstock CL, Helman WP, Ross AB. Critical review of rate constants for reactions of hydrated electrons, hydrogen atoms and hydroxyl radicals in aqueous solutions. J. Phys. Chem. Ref. Data. 1988;17(2):513–886. www.rcdc.nd.edu/compilations/Eaq/Index.html.

[10] Galanin MD. On the reasons for the dependence of luminescence yield of organic chemicals on the energy of ionizing particles. Optics Spectroscopy 1958;4(6):759–62.

[11] Kabakchi SA, Bulgakova GP. Radiation chemistry in the nuclear fuel cycle. Tutorial - http://www.chemnet.ru/rus/teaching/kabakchi/welcome.html#2 – in Russian.

[12] Halliwell B. Free radicals in biology and medicine Halliwell B, Gutteridge JMC.- 4th ed., Oxford Univ. Press, 2007.- 851.

[13] Le Caër S, Water Radiolysis. Influence of oxide surface on H_2 production under ionizing radiation. Review. Water 2011;3:235–53. http://www.mdpi.com/journal/water.

[14] Timofeeff-Ressovsky NV, Ivanov VI, and Korogodin VI . Application of the "Hit" Principle in Radiobiology. Atomizdat, 1968, 228 p. [In Russian.(Тимофеев-Ресовский Н.В., Иванов В.И., Корогодин В.И. Применение принципа попадания в радиобиологии. М., 1968. 228 с.)]

[15] Lea DE. Action of radiation on living cells. Cambridge, MA: Cambridge University Press; 1947.

[16] Eidus LKh, Eidus VL. Problems of the Mechanism of Radiation and Chemical Hormesis. Radiation Biology. Radioecology. 41, No. 5 (2001) 627-630. [In Russian.(Л.Х. Эйдус и В.Л. Эйдус. Проблемы механизма радиационного и химического гормезиса. Радиационная биология. Радиоэкология. 2001, т. 41, №5, 627-630)]

[17] Yu.B. Kudryashov. Radiation Biophysics (ionizing radiation). A textbook. Fizmatlit, Moscow, 2004, 443 p. [in Russian (Кудряшов Ю.Б. Радиационная биофизика (ионизирующие излучения). Учебник. М.: Физматлит, 2004. – 443 с.]

[18] MotherSill C, Seymour C. Changing paradigms in radiobiology. Mutation Res. 2012;750:85–95.

[19] Blyth BJ, Sykes PJ. Radiation-induced Bystander effects: what are they, and how relevant are they to human radiation exposures? Radiation Res 2011;176(2):139–57.

[20] US Department of Energy. Office of Science. Low Dose Radiation Research Program. Research Highlights. Radiation-Induced Bystander Effects and relevance to Human Radiation Exposures - http://lowdose.energy.gov/radiation_bystandereffects.aspx.

[21] G. Rattue. Abscopal Effect - When Radiation Also Destroys Non Targeted Tumors, 2012 - http://www.medicalnewstoday.com/articles/242784.php.

[22] UNSCEAR Report 2006, Annex C. Non-targeted and delayed effects of exposure to ionizing radiation - http://www.unscear.org/docs/reports/2006/09-81160_Report_Annex_C_2006_Web.pdf.

[23] A.V. Shafirkin, V.V. Bengin. The peculiarities of the action of cosmic radiation on biological objects and radiation risk of long-term space missions. Institute of Nuclear Science of MSU - http://lib.qserty.ru/static/tutorials/01_textbook/index-908.htm - in Russian.

[24] Atomic bomb victims' appeal. - http://www.ne.jp/asahi/hidankyo/nihon/rn_page/english/1988.htm.

[25] Rajewski B, Heuse O, Aurand K. Weitere Untersuchungen zum Problem der Ganzkorper Bestrahlung der weissen Maus. Sofortiger Tod durch Strahlung. Zeitschrift Naturf 1953;86:157–9.

[26] Feldman R. Radiation injury. In: Schaider JJ, Hayden SR, Wolfe RE, Barkin RM, Rosen P, editors. Rosen and Barkin's 5-Minute Emergency Medicine Consult. 3rd ed. Philadelphia, PA: Lippincott Williams & Wilkins; 2007.

[27] Thierry de Revel, Menace terroriste approche medicale. Nucleaire Radiologique Chimique, 2005.

[28] Tanner AB. Radon migration in the ground: a review The Natural Radiation Environment. In: Adams JAS, Lowder WM, editors. University of Chicago Press: Chicago, IL; 1964. p. 161–90.

[29] Tanner, A.B. Radon migration in the ground: a supplementary review. 5-56 in: Natural Radiation Environment III, Volume 1 (T.F. Gesell and W.M. Lowder, eds.).

[30] UNSCEAR 2000 Report. Annex B. Exposures from natural radiation sources. II. Terrestrial Radiation. C. Radon and decay products. 96-108 - http://www.unscear.org/unscear/publications/2000_1.html.

[31] UNSCEAR 1993 Report. Annex A: Exposures from natural sources of radiation - http://www.unscear.org/unscear/en/publications/1993.html.

[32] ICRP, 2010. Lung Cancer Risk from Radon and Progeny and Statement on Radon. ICRP Publication 115, Ann. ICRP 40(1).

[33] Cohen BL. Test of the linear no-threshold theory of radiation carcinogenesis for inhaled radon decay products. Health Phys 1995;68:157–74.

[34] Cohen BL. Problems in the radon vs. lung cancer test of the linear no-threshold theory and a procedure for resolving them. Health Phy. 1997;72:623–8.

[35] Cohen BL. Radon in air. In: Lehr JH, Lehr JK, editors. Standard Handbook of Environmental Science, Health and Technology. New York, NY: McGraw-Hill; 2000. 15.7–15.19.

[36] Cohen BL. Review: Cancer risk from low-level radiation. Am. J. Roentgenol. 179, No. 5 (2002) 1137–43.

5 The Effect of Chemicals on Biological Structures

5.1 Chemicals

5.1.1 The Registers of Chemicals

How many substances are there in the world? Let us refer to the official registers of chemicals. There are several such registers (databases). We are going to use one of the most authoritative databases of chemicals, the Chemical Abstracts Service (CAS [1]). The register of CAS of mid-August 2014 contains information on nearly 90 million chemicals. The exact figure is constantly changing. Two years ago (mid-July 2012), it was nearly 70 million chemicals.

Modern rate of adding new chemicals into the database is approximately 200,000 per week (~20 per minute).

The well-known natural substances were registered in the first years after establishing of this service in 1907. Now mainly the recently discovered or synthesized and patented compounds are recorded.

According to the European Commission's White Paper on the Strategy for a Future Chemicals Policy (February 2001) [2], the world production of chemicals has increased from 1 million tons in 1930 to 400 million tons in 2001. The Chemical Industry of the European Union (EU) is one third of the world chemical production. About 100,000 different substances are registered in the EU, of which more than 10,000 appear on the market each in the volume of more than 10 tons and 20,000 in the volume of 1–10 tons each.

Furthermore, there are huge numbers of substances produced uncontrolled in production process and in cooking. For example, in the process of roasting coffee, about a thousand different substances are formed. Some of the substances appeared this way are not yet even identified. Some substances from this diversity can be hazardous, harmful to health.

Nowadays, humanity has to deal with thousands of toxic substances. Information on these substances, their chemical and physical properties, and their toxicity can be found in databases of toxic substances [3–5].

For a detailed reference of the databases, see Section 5.4.1.

The extensive summary data can be found on the site of the U.S. National Institutes of Health [6]. The Hazardous Substances Data Bank (HSDB) [7] offers information on more than 5725 potentially dangerous substances (August 2014). The International Toxicity Estimates for Risk (ITER) reports about more than 650 hazardous substances [8]. An online version of the database of the Russian Automated Distributed Data Retrieval System, "Hazardous Substances," by the Russian Register of Potentially Hazardous Chemical and

Biological Substances of Russian Agency for Health and Consumer Rights (Rospotrebnadzor) [5]—as of July 23, 2014, contains information on about 9586 substances. The Carcinogenic Potency Database (CPDB) [9] specializes in carcinogens, and it reports about carcinogenicity testing of more 6540 substances.

The results of testing show that not all of the substances in these databases turned out to be carcinogens, but about half of them are really dangerous.

Compare the above-mentioned numbers with the number of chemicals in the CAS register. It is evident that people have to deal with a huge number of different substances, which risk level is not yet known, and a large number of really dangerous substances.

5.1.2 Intensive Chemistry Development

Although the development of chemistry began long before the Christian era, the stage of wide practical application of chemistry in many industries should be approximately attributed to the middle of the eighteenth century. It coincided with the industrial revolution. Intensive mining, development of metallurgy, power engineering, and transport caused enormous growth of production forces, but it also turned out to be a substantial burden for the environment. The second half of the twentieth century marked a new stage in the development of chemistry.

Many scientific advances, coming to everyday life, were associated with chemistry: synthetic textiles, synthetic plastics, fertilizers, and pesticides. Synthetic clothing from nylon, acetate, polyamide, polyester, and other synthetic materials has now become customary and comfortable.

The chemical industry, a complex conglomerate of various industries with a wide range of products, is one of the main suppliers of pollutants to the environment.

First, the chemical industry is harmful to its employees. This is considered by the special working conditions for employees in "harmful" industries. In hazardous industries and occupations with severe and harmful working conditions, the employment of women is prohibited. Occupation in hazardous industries offers some special privileges.

Harmful production factors can lead to a decrease of labor efficiency, occupational diseases and negative long-term effects. The list of hazardous occupations includes, of course, some professions associated with other types of hazards, such as possible electrical accidents or work in hot shops, manual labor, work with pneumatic tools, and so on; however, the vast majority of hazardous working conditions are present in the chemical plants.

But the chemical industry is harmful not only for the employees but also for the environment and the whole population. That can be due to the low-quality functioning of filter units and purification plants, which result in a lot of waste, the storage and processing of which has not yet adequately advanced in some countries. The greatest harm to the environment and the population is brought about by various accidents in the chemical industries, the chemical spills, in particular.

The spills may appear at different plants, during the transportation, due to accidents, and due to noncompliance with safety regulations. There are a lot of such instances.

Among the various incidents, there are some very exotic ones. For instance, in early August 2003 in Qiqihar (Heilongjiang Province, in the northeast of the People's Republic of China) the workers on a construction site found some metal vessels. They cut those vessels into parts and sold them as scrap metal at one of recycling terminals in the town's residential district. The find, as it turned out, were the empty mustard gas vessels from a Japanese chemical weapon stock abandoned in World War II (1937–1945). As a result, 43 people were poisoned with varying severity. The soil at the site where the vessels were found was contaminated.

At 9 a.m. on July 27 at a plant for chemical weapons destruction at the military unit 21225 "Pochep-2" in the Bryansk region, a leakage of 6 tons of the nerve agent VX was detected. VX is a chemical warfare, an organophosphorus nerve agent. Gassing occurs through the respiratory tract and skin. The symptoms appear as follows: in 1–2 minutes, constriction of the pupils; in 2–4 minutes, sweating, salivation; in 5–10 minutes, convulsions, paralysis, spasms. Death from gassing with VX occurs within 10–15 minutes.

Most recently, people have been faced with a new problem that they never had to deal with before, namely, the toxic and hazardous nanoparticles and nanomaterials. As a matter of fact, fine dust already existed long before the formation of "nanotechnology" as an independent phenomenon, though in smaller amounts and was not purposely produced.

Because of their small size, nanoparticles can penetrate unchanged into the body, breaking through the various barriers (blood–brain, placental barriers, etc.) that nature has built to resist foreign substances, getting through the skin, the respiratory and gastrointestinal epithelium, and then storing up in the bone marrow, central and peripheral nervous system, gastrointestinal tract, lungs, liver, kidneys, and lymph nodes. Nanoparticles have a prolonged half-life. A new branch of research, nanotoxicology, has appeared. It is essential to study the interaction of nanostructures with biological systems in order to identify the connection between physical and chemical properties of nanomaterials (such as size, shape, surface properties, their composition, and the degree of aggregation) with induction of the toxic response in biological structures.

Thus, the main source of environmental pollution is the chemical industry and especially the various industrial accidents as well as those during transportation of chemical products. However, the author believes that the vast majority of people, and in particular the readers of this book, are most likely not going to get into the zone of an accident and they do not work in hazardous industries. But, unfortunately, they are still forced to have contact with harmful substances in the air, in water, in food, in detergents, and in products of personal care.

5.1.3 The Air

5.1.3.1 The Outdoor Air

The Earth's atmosphere consists of nitrogen, oxygen, argon, carbon dioxide, and water vapor. Concentration of oxygen and nitrogen has been constant for many thousands of years: dry air contains 78.084% of nitrogen and 20.95% of oxygen (volume percent). In the atmosphere, there is quite a lot of argon (0.93%), mainly its isotope ^{40}Ar produced during the decay of the radioactive nuclide ^{40}K (see Section 1.10.2).

Standard atmosphere also contains carbon dioxide, CO_2 (0.0395% = 395 ppm in dry air, as of 2014). Measurements show that currently the atmospheric concentration of CO_2 is rising at the rate of 2.20 ± 0.01 ppm/year and accelerating. It is assumed that the increase in carbon dioxide concentration could pave the way for increasing of the role of the greenhouse effect, for increasing of the average temperature on earth, for the glacier melting, for the rise of the ocean level, and other unpleasant consequences, all hazardous for human health.

These figures above relate to the dry air. But in reality, water vapors are always present in the air. The water vapor concentration substantially depends on the temperature (vapor pressure) and on humidity. It is believed that the average proportion of water vapor in the atmosphere is about 1%. However, in the hot humid tropics, its concentration can reach up to 7%.

The other gases in the atmosphere are inert gases (helium, neon, krypton, and xenon), hydrogen, methane, nitrogen oxides, carbon monoxide, ammonia, sulfur oxides, hydrogen sulfide, and hydrocarbons. The share of these substances in a "clean" atmosphere is only about 0.0004%.

An interesting, though not very well known, fact is that many gases, including nitrogen, hydrogen, and nitrogen oxides and all inert gases, can have a narcotic effect when they penetrate into the lipid layer of the cell in sufficient concentration. Most of the "inert" gases exhibit anesthetic properties only under higher pressure. For example, with the diving community, it is assumed that nitrogen becomes a dangerous narcotic at a depth of 30–40 m under water (i.e., at a pressure of 4–5 absolute atmospheres). With a little more pressure, narcotic properties are shown by argon and hydrogen. Helium and neon are considered nonnarcotic gases because of their small (but not zero) narcotic ability (see Table 5.1). Only xenon and, to a lesser extent, krypton become drugs at atmospheric pressure [10].

Narcotic abilities of different gases are shown in Table 5.1. This is what the method of inhaled xenon anesthesia is based on. A standard inhalation mixture composition is 70% of xenon and 30% of oxygen.

Table 5.1 Narcotic Effectiveness of Certain Gases and Their Solubility in Lipids

Gas	Solubility in Lipids	Relative Narcotic Effectiveness
Helium	0.015	0.2
Neon	0.019	0.3
Hydrogen	0.036	0.6
Nitrogen	0.067	1.0
Argon	0.140	2.3
Krypton	0.430	7.1
Xenon	1.700	25.6

Xenon anesthesia doesn't have any of the drawbacks of all other methods. There are no side effects, xenon gets very rapidly excreted out of the body, and the patient regains consciousness almost immediately after the cessation of inhalation. The exact mechanism of the narcotic effect of xenon is not yet fully clear, but it is clear that it is fundamentally different from the actions of other anesthetics. Apparently, xenon dissolves in the lipid layer of nerve cell membranes and changes nerve conduction. Specialists claim that xenon's action has a mechanical effect rather than chemical [11]. The narcotic ability of gases correlates well with their solubility in lipids, as is clear from Table 5.1.

Narcotic ability is also demonstrated by nitrous oxide (nitric oxide I [N_2O]). Small concentrations of nitrous oxide cause a feeling of intoxication (hence the name "laughing gas") and light drowsiness. At high concentrations, it is used in inhalation anesthesia. However, all nitrogen oxides influence the human nervous system. Nitric oxide (nitric oxide II [NO]) causes paralysis and convulsions, binds hemoglobin, and causes oxygen starvation. Nitrogen dioxide (nitric oxide IV [NO_2])—causes lesion of the respiratory tract and pulmonary edema.

In August/September 2012, a wave of comments went through Russian media about a fascination with inhaling nitrous oxide at discos, which was widely spread among young people. The gas was sold in ordinary kids' balloons. The same situation was in the world.

Inhaling nitrous oxide "to get high" is freely advertised on the Internet. For some consumers (the main buyers were pupils and students), the moments of fun ended dramatically—with mental disorders, hypoxia, and even death. Statistics confirm that the laughing gas market is growing up to now [12].

Among the main suppliers of pollution into the atmosphere are volcanoes. There are about 10,000 volcanoes on Earth, most of them are considered "dormant" since intervals between their active periods may far exceed the history of humanity. Nevertheless, nearly 500 terrestrial volcanoes are considered active. The largest emission during eruptions occurred 74,000 years ago from the Toba volcano in New Zealand. According to the estimates, 2800 km^3 of a matter was thrown out. According to one of the theories, the eruption of this volcano caused a significant reduction in the world's population, the ancestors of modern humans. Apparently, several grand eruptions are associated with the volcano of Santorini on the island of the same name in the Aegean Sea. An eruption occurred about 3500 years ago, which led to the end of the Cretan–Mycenaean civilization and possibly ruined the legendary Atlantis. One of the three active volcanoes in Italy, Vesuvius, which erupted in 79 BC and buried the towns of Pompeii and Herculaneum, threw out $\sim 1\ km^3$ of a matter. A relatively recent eruption of the Krakatoa volcano in Indonesia in 1883 led to the release into the atmosphere of about 150 billion tons of dust and ash. Eruptions of our time have also contributed to significant air pollution: St. Helena (Washington) in 1980, $\sim 1\ km^3$ of ash and dust; Pinatubo (on the island of Luzon in the Philippines) in 1991, $\sim 10\ km^3$ of matter. In April 2010, the Icelandic volcano Eyjafjallajökull threw "only" $<0.01\ km^3$ of matter, but because of this, part of Europe's airspace was completely closed for 5 days, and occasional flight restrictions took place throughout May 2010.

Volcanoes emit oxides of sulfur and chlorine and some other substances, both separate and within the composition of solids, ash, and dust. Thus, an eruption in 1976 of the volcano Augustine in Alaska emitted 300 million tons of magma containing up to 0.5 million tons of chlorine; Etna in Sicily emits more sulfur dioxide than 50 large coal-fired power plants. Once in the atmosphere, sulfur dioxide gets very quickly converted into sulfuric acid. Chlorine is ejected mainly in conjunction with hydrogen, that is, in the form of hydrogen chloride. In the lower layers of the atmosphere, it rapidly reacts with water, forming hydrochloric acid.

Other natural phenomena that lead to air pollution are forest fires and dust storms. During strong dust storms, up to 50 million tons of dust rises into the air.

Catastrophic volcanic eruptions occur rarely and irregularly, other natural sources contribute much less to pollution. But artificial (anthropogenic) sources operate continuously and since some time their activity has been growing.

According to data cited by the Web-Atlas, "Environment and the Health of Russian population," 20.2 million tons of harmful substances were thrown into the atmosphere in 1996 by Russian enterprises. About a third of this amount is accounted for by sulfur dioxide, and about a quarter by carbon monoxide. The rest (in descending order) is solid matter, nitrogen oxides, hydrocarbons, and volatile organic compounds. The great bulk of emissions (85%) is produced by industry. Agriculture and public sector account for only 15% of emission volume. In industry, the major air pollutant is the electric power sector, which accounts for about a quarter of the total volume of emissions in the country, and together with the fuel industry, more than 40%. Another 30% is produced by the metallurgy (nonferrous metallurgy produces more emission than iron) industry. Other sectors taken together make up 15% [13].

According to Russian Hydrometeorology and Environmental Monitoring Agency (RosHydroMet) [14], at least 4.22 million tons of sulfur and 4.0 million tons of nitrogen (nitrate and ammonium) annually fall out in Russia in the form of acidic compounds in precipitation. The biggest fallouts of sulfur are observed in densely populated and industrialized regions. Incidentally, a significant proportion of pollution is associated with cross-border transfer. Some emissions of nitrogen oxide are transformed into nitrogen dioxide. At low concentrations of nitrogen dioxide, impaired breathing and coughing may occur. WHO recommended not exceeding the level of 400 $\mu g/m^3$ in order to prevent painful symptoms in patients with asthma and other groups of people with high sensitivity. With an average annual concentration of 30 $\mu g/m^3$, the number of children with hyperventilation, cough, and bronchitis increases.

In the United States, air pollution causes about 200,000 early deaths each year, according to a study from the Massachusetts Institute of Technology (MIT) [15]. Six major air pollutants—carbon monoxide, ground-level ozone, lead, nitrogen dioxide, particulate matter, and sulfur dioxide—are monitored. It is estimated that the amount of carbon monoxide pumped into the atmosphere throughout the United States every year is on the order of nearly 60 million tons. The percentage of lead air pollution has fallen with the greatest speed and effectiveness of all major pollutants. From 1980 through 2010, the level of lead polluting air in the United States dropped by 89% [16]. The official information about the air quality can be found in the site of the U.S. Environmental Protection Agency [17,18].

As shown in [13], average concentrations of nitrogen dioxide increase markedly from north to south due to the influence of solar radiation on photochemical reactions of nitrogen oxide transition into dioxide.

Among the pollutants released into the atmosphere due to man-made emissions from industry, power plants, and transport, nitrogen oxides deserve special consideration. They are formed mainly at the high temperature combustion of fossil fuels in the form of nitrogen oxides which are getting transformed into NO_2. All emission evaluations are usually expressed as compared to NO_2, although it is impossible to determine exactly what proportion of emissions presents in the atmosphere in the form of NO_2 or NO. Nitrogen oxide and dioxide play a complex and important role in photochemical processes that take place in the troposphere and stratosphere under the influence of solar radiation, causing the formation of photochemical smog and high concentration of ozone O_3.

Urban residents are familiar with the exhausts of car engines (both petrol and diesel).

For the urban areas of Russia, rather high concentrations of formaldehyde are typical. The average level for the cities is 8 $\mu g/m^3$, which is almost three times higher than the maximum allowable concentration (MAC).

The Environmental Monitoring Systems in Russia takes stock of the following impurities in the atmosphere: sulfur dioxide, SO_2; carbon monoxide, CO; nitrogen dioxide, NO_2; nitric oxide, NO; benzopyrene, BP; formaldehyde, F; suspended solids, SS, ethyl benzene, EB; carbon sulfide, CS_2; and ammonia, NH_3. Also, such relatively rare pollutants as hydrogen fluoride HF and methyl mercaptan are monitored.

For the cities with a population of less than 250,000 people, the following standards of background concentrations of the major toxicants are accepted:

$SO_2 = 0.1$ mg/m^3, CO = 1.5 mg/m^3; $NO_2 = 0.03$ mg/m^3; dust = 0.2 mg/m^3

The main air pollutants, 25 billion tons of which are emitted annually, are the following:

- sulfur dioxide and dust particles = 200 million tons/year;
- oxides of nitrogen (N_xO_y) = 60 million tons/year;
- carbon oxides (CO and CO_2) = 8000 million tons/year;
- hydrocarbons (C_xH_y) = 80 million tons/year.

An anthropogenic stream of toxicants released into the environment mostly prevails over the natural (50%–80%) and only in some cases is comparable to it. Volatile organic compounds released into the atmosphere quite easily enter into chemical reactions with oxygen and other oxidants, which leads to the formation of stronger toxic pollutants.

It has to be noted that efforts to reduce emissions and purify the atmosphere are being made worldwide and, of course, they bring results. Well known are the famous smogs, which used to be almost normal for London, Los Angeles, and many other cities. In 1952, in London, 2500 people died from smog within a few days. In Tokyo in 1973, smog stayed for 328 days. Tokyo traffic police were on duty in oxygen masks, and slot machines selling servings of clean air were installed in the streets. But by now the danger of smog has virtually disappeared. The requirements for automobile exhaust purification systems are continuously tightened. According to the latest requirements, the exhaust must contain an average of less than 100 g of carbon dioxide per kilometer of mileage.

5.1.3.2 The Indoor Air

The indoor air composition largely repeats the composition of the surrounding atmosphere. However, there may be significant differences. Concentrations of various substances in a room may be higher than the outside, since a lack of wind allows pollutants to accumulate. In an unventilated room, especially with a lot of people in it, the concentration of carbon dioxide grows up to 0.3%–0.5%, and sometimes even up to 0.8%; that is, it can grow by more than 20 times! Besides, while breathing, people exhale carbon monoxide, methane, ammonia, aldehydes, and ketones. With the help of gas and liquid chromatography methods, Linus Pauling identified up to 250 substances in exhaled breath condensate [19], and modern techniques enable the identification of up to 1000 substances [20].

The indoor air can contain phenol and formaldehyde from evaporations produced by modern cheap furniture, ammonia from concrete and masonry mortar in some houses, and tobacco smoke, which includes hundreds of organic compounds, dust, aerosols, and many other, often very harmful, components.

5.1.4 Water

We are not going to mention here the microbiological contamination of water or air, since the purpose of this book is to analyze the impact on health of chemical and radiation factors. However, one must understand that both water and air may contain viruses, bacteria, fungi and mold spores, and, besides, in the air are also present dust mites and saprophytes. Besides, it is often difficult to divide pollutions into components.

Pollution of surface waters began in central Russia as early as the sixteenth century, when people started to fertilize fields with manure. Since then, agriculture has been the main polluter of water in central parts of the country. Further to the North, timber rafting played a major role, especially loose floating when a lot of logs were drowned and rotted in the water. With the development of industry and urban growth, the role of municipal and industrial pollution began to increase.

The main substances that pollute surface water are oil products, phenols, easily oxidized organic matters, copper and zinc compounds, ammonium and nitrates. For several decades because of industrial and municipal discharges one cannot drink the water from the Middle and the Lower Volga [13].

Currently, according to the Russian Register of Potentially Hazardous Chemical and Biological Substances of Rospotrebnadzor (Russian Agency for Health and Consumer Rights) [21], up to 52 km^3 of waste water per year is discharged into water bodies of the Russian Federation, of which 19.2 km^3 are subject to treatment. More than 72% of waste water to be treated (13.8 km^3) is discharged into water bodies insufficiently treated, 17% (3.4 km^3) contaminated and without treatment, and only 11% (2 km^3) are treated according to established standards. Along with wastewater, surface water bodies of the Russian Federation annually receive about 11 million tons of pollutants.

Main sources of contaminated waste water are housing and communal services, industry, and agriculture. These account for more than 90% of the total volume of contaminated waste water discharge.

Domestic runoff water contains chemicals and synthetic fertilizers and detergents. Emissions from industrial effluents contain organic and nonorganic substances that are by-products of industrial production.

Oversaturation of stagnant or slow water with nitrates and phosphates leads to eutrophication (i.e., rapid growth and reproduction of microorganisms and algae).

Enormous damage to the environment and, in particular, to water basins is caused by oil spills. Oil spilt from damaged pipelines, tankers, and offshore installations destroys all life in their working area. Oil spreads over the surface of the water for many miles, and when it reaches the shoreline, it tightly sticks to sand, rocks, and stones. Because of oil pollution, all vegetation gets killed. Contaminated areas become unsuitable for wildlife habitat.

Some oil particles can be mixed with water and sink down to the bottom, thereby killing the sensitive marine ecosystem. Many marine organisms and fish die or get infected.

Thus, for example, in 1989 there was an enormous oil leakage in Alaska. Despite attempts to eliminate the effects, the probes taken in 2007 showed that even 18 years after the accident, 26,000 gallons of oil still stayed in the sand along the shoreline. Naturally, the wildlife populations in these areas have not recovered.

It was determined that after elimination of spills, oil residue are disappearing with a rate of 4% of the total oil volume per annum.

On April 20, 2010, in about 80 km off the Louisiana coast of the Gulf of Mexico, an accident with subsequent oil spill occurred on the oil rig Deepwater Horizon (the oil field of Macondo). Through damaged pipes of the well at a depth of 1500 m, nearly 5 million barrels of oil spilled into the Gulf of Mexico in 152 days (about 800 million l = 800,000 m^3). The oil slick spread over an area of 75,000 km^2. Spilled oil is causing the death of fish, birds, and marine mammals. As of November 2, 2010, a total of 6814 dead animals were found.

Actually, all kinds of water pollution can be dangerous to humans and animals. Contaminated water does not cause an immediate harm to human health, but can gradually undermine the health. Various kinds of pollution act differently.

Heavy metals can accumulate in aquatic plants, can get into a fish, and into a human who eats the fish. Teratogenic and carcinogenic effects may be exhibited.

Industrial wastes often contain toxic components that also affect the health of inhabitants of rivers and lakes.

Organic substances and nitrates promote the growth of oxygen-consuming flora and thus reduce the oxygen content in water. The life of aquatic species can be dramatically influenced by sulfates from acid rains. Suspended solids while getting from the atmosphere into water impair the quality of drinking water for humans and the conditions necessary for aquatic life. Suspended particles make water turbid, reduce the amount of sunlight penetrating the water, and reduce opportunities for growth of photosynthetic algae and microorganisms.

Here are two examples of pollution of water basins in foreign countries, the first, Lake Erie, one of the Great Lakes of North America, the second, the Rhine River in Europe.

5.1.4.1 Lake Erie

By the mid-twentieth century, anthropogenic impacts caused severe pollution of the Great Lakes. Annually, 46 million tons of solids, 4 million tons of chlorides, 27,000 tons of phosphates, and 160,000 tons of nitrogen products were discharged into one of the Great Lakes—Lake Erie. Pollution of the lake did not draw much attention until the Great Fire, which happened on the Cuyahoga River in June 1969. The Cuyahoga River that flows through the city of Cleveland at the end of the 1960s was so badly polluted with oil products that sometimes it flamed up. Along with it, increasing levels of phosphates in water and bottom deposits caused an eutrophication of the whole ecosystem, because of fundamental changes in algae productivity and blooms. Decomposition of algae provoked an expansion of seasonal dead zones in the lake. Masses of decaying algae and polluted coastlines helped to spread the rumors of Lake Erie as a dead lake. People said that "it has become too thick to swim in, but still too liquid to plough." In recent decades, measures have been taken that have improved the ecological situation, although it is still far from ideal.

5.1.4.2 The Rhine—The Former Gutter of Europe

Contamination of the Rhine, associated with industrialization, began in the second half of the nineteenth century and reached its peak in the early 1970s. On the Rhine and its tributaries, a large number of industrial enterprises of Switzerland, France, Germany, and the Netherlands were established. Also, intensive hydraulic reorganizations were conducted—the river straightening and construction of hydroelectric power plants. Consequently, water quality became so bad that the Rhine became known as a gutter or a cesspool of Europe. Back in 1950, the International Commission for the Protection of the Rhine developed the action plan for improving the Rhine ecosystem. The last event resulting in dramatic consequences for the river was the great fire that happened in 1986 at the chemical plant near Basel (Switzerland), when 30 tons of chemicals—pesticides, mercury, and other agricultural chemicals—got into the Rhine. The Rhine water turned red. People who lived by the river were forbidden to move out of their homes, the public water supply in some German cities was stopped. Within 10 days, the water pollution reached the North Sea. All this led to mass death of fish, some species disappeared altogether. After a violent public reaction in 1987, the most stringent measures for environmental protection in the scale of large industry were introduced. A special program of actions right up to 2000 was adopted. One of the suggested titles for the program was "Salmon 2000." The salmon returned to the river in 1997, 3 years before the deadline.

In 2001, 31 species of fish were registered in the Rhine, including salmon and herring, which had been considered extinct in the river. Improved water quality had a favorable impact on other species of living organisms, and currently the diversity of animal species in the Rhine has almost returned to the condition at the beginning of the twentieth century.

It is worth noting that as a result of aggregate measures to protect the environment, not only for the Rhine River but also for other regions of Germany, an environmental industry with enormous financial turnover was established that has a greater number of employees than in the car industry.

In order to eliminate pathogenic bacteria, especially typhoid, cholera, and dysentery, tap water is chlorinated by municipal purification plants. The sterilizing effect of chlorine is caused not quite by Cl_2 but by the hypochlorous acid (HOCl) produced by the reaction of chlorine with water. Although chlorine for many years was used to sterilize water without causing significant deleterious effects to human health, it was recently discovered that it can, in the long run, inflict some damage. Combining with possible impurities of organic compounds, chlorine forms so-called halomethanes, for example, chloroform ($CHCl_3$) or carbon tetrachloride (CCl_4). The toxicity of these substances does not admit any doubt.

5.1.5 Food

Various chemicals that may be present in food should be divided into two groups: substances that are purposely introduced into food, so-called food supplements, and substances that are formed in the food during cooking. Some of these substances can directly affect health. Of course, in most cases, except those obviously criminal, these substances are not direct toxicants, but they can have long-term effects, mainly on the genetic apparatus. Some products of a cooking process can turn out to be carcinogenic.

Currently, many food producers add various dietary supplements into food products. Actually, the various "improvers" of food have been known since the ancient times. But in former times, people used natural products. Now increasingly more synthetic chemicals are introduced and the chemical industry is constantly developing. It should be expected that the growth of synthetic additives will continue.

The food industry uses a large number of different food additives [22]. All of them are marked with the letter E and a number. The food supplements family consists of colorants (E100–E199), preservatives (E200–E299), antioxidants (E300–E399), stabilizers, thickeners, emulsifiers (E400–E499), raising agents (E500–E599), flavor enhancers (E600–E699), antibiotics (E700–E799), coating agents, food sealers, separators, and propellants (E900–E999). For other types of additives that may arise in the future, additional codes E1000–E1999 are secured. Some supplements are neutral, some are harmful to health, and some are dangerous. Obviously, dangerous supplements are banned, but in some cases these prohibitions have been issued only recently and foods produced prior to issuance of the relevant law might contain those supplements. Besides, the regulations on the use of nutritional supplements, say, in the Russian Federation may differ from those of some other countries.

Let us mention some specific chemical substances used as food additives. Preservatives: E210–E214, benzoic acid and benzoates (of sodium, potassium, and calcium), are carcinogens; E219, methyl paraben, is a carcinogen; E222 and E223, sodium hydrosulfite and pyrosulfite, are harmful to health; and E281–E283, propionates, are carcinogens. Antioxidants: E313, ethyl gallate, is banned; E324, ethoxyquin, is banned. Of all flavor enhancers, glutamic acid and glutamates, the salts of this acid, only sodium glutamate (E621) is allowed for use. According to the information on the Consumer Rights Protection Society, see [23], glutamic acid (E620) is dangerous. Other glutamates, the additives E622–E625, are not allowed for use in the Russian Federation.

On the issue of safety or health hazards of food additives, it is essential to note that many amplifiers and taste modifiers, perhaps, originally are safe, but they are often added to conceal the poor quality and nutritional value of products. Such amplifiers can be found in practically all fish, chicken, mushroom, and soy half-stuffed foods, as well as in chips, crackers, sauces, various dry spices and soups, and bouillon cubes.

A list of food additives that are not approved for use in the Russian Federation, as well as a list of dangerous and even hazardous (but allowed) additives causing gastrointestinal disorders, blood pressure disorders, rash, and other unpleasant consequences, can be found on the Consumer Rights Protection Society site [23].

In spite of the fact that the United States is the most developed country in the world, they have a lot of problems with the quality of food as well. As it is said on the site of the Food and Drug Administration (FDA), "About 48 million people (one in six Americans) get sick, 128,000 are hospitalized, and 3000 die each year from foodborne diseases, according to recent data from the Centers for Disease Control and Prevention. This is a significant public health burden that is largely preventable." [24]. The authors of the book "Rich Food Poor Food," M. Calton and J. Calton, wrote, "For numerous suspicious and disturbing reasons, the U.S. has allowed foods that are banned in many other developed countries into our food supply" [25]. The possibility to use in the United States the products that are banned in many countries is actively discussed in the press [26].

The FDA has a list of food additives that are considered safe. Many have not been tested, but they are considered safe by most scientists. These substances are put on the "generally recognized as safe (GRAS)" list. The list contains about 700 items. Some substances that are found to be harmful to people or animals may still be allowed, but only at the level of 1/100th of the amount that is considered harmful [22,27].

It is important to understand that in food during or after it is processed, substances can be found that were not used or placed in the food on purpose. Such substances are called indirect food additives [22].

5.1.6 Detergents, Cosmetics, and Personal Hygiene Products

Detergents have washing properties due to superficially active substances (SAS) that make up the bulk of all detergents, that is, shampoos, liquid soaps, shower gels, bubble baths, and so on.

SAS are complex organic compounds, derivatives of hydrocarbons. They are composed of a polar hydrophilic and a nonpolar (hydrocarbon) hydrophobic portion. Surface activity in relation to a nonpolar phase (the surface of a solid body) is a quality of a hydrocarbon radical. Carboxylic acid derivates, alkyl sulfates, ester sulfonates, and many other substances can be used as SAS. The presence of complex organic surfactants, cyclical components in particular, can make them toxic to humans. Although, as a rule, this toxicity is slightly expressed, it may appear because of the frequency and intensity of people's contact with detergents. Furthermore, the waste water that contains surfactants may have an impact on the environment and the living organisms. Therefore, great attention must be paid to the clearing of the waste water of surfactants.

Surfactants may not only cause a general toxic and allergenic effect on the human body but also enhance the toxicity and carcinogenic, sensitizing, and mutagenic effect of other chemicals.

Another source of the negative influence of detergents on human health is the presence of phosphorous compounds in their structure. Phosphorous compounds are used to soften water and to enhance the cleaning properties of powders. Phosphorous compounds penetrate into the skin cells and may cause skin diseases. Furthermore, phosphorous compounds may directly penetrate into the blood and change its composition. Dysfunctions of liver, kidneys, and skeletal muscles may occur, which in turn leads to severe poisoning, metabolic disorders, and exacerbation of chronic diseases. It has been proved that the major mechanism of phosphorus compound action is their interaction with the lipid-protein membranes and penetration through them into the various structural elements of the cell, thereby causing slight changes deep in the biophysical and biochemical processes.

Phosphates are difficult to rinse out of washed clothes. Phosphorous compounds make a surfactant almost irremovable. In order to remove any residual phosphates, more than 10 rinses under running water are needed. On average, potentially unsafe concentrations of phosphates are retained by tissues for up to 4 days. Firmly entrenched on clothing, phosphates getting in contact with skin relatively easily transfer to its surface and are rapidly absorbed.

Phosphates are also extremely harmful for the environment. For example, getting into the water, these compounds fertilize algae so generously that dead-water blooms abundantly and turns marshy.

In developed countries, phosphate detergents are either banned or their use is fully restricted.

5.1.6.1 Harmful Substances in Cosmetics

As reported by the organizations involved in health protection, more than 10,000 different substances are used as personal care products. Some of these substances may be carcinogenic, causing problems in childbirth, disrupting reproductive capabilities, or causing some other harm. The U.S. FDA [28] banned the use of nine ingredients in cosmetic products. In the European Union, more than 1000 ingredients are banned [29]. Harmful and hazardous substances may be found in different cosmetics (creams, lotions, toothpastes, hair dyes, shaving cream, etc.). Lists of these substances are easy to find online.

As reported on the FDA site, substances prohibited for use are the following: bithionol can cause photocontact sensitivity; chlorofluorocarbon propellents have a harmful impact on the atmosphere; halogenated salicylanilides: di-, tri-, metabromsalan and tetrachlorosalicylanilide can cause photocontact sensitivity; chloroform, methylene chloride, and vinyl chloride are potential carcinogens; and zirconium-containing complexes have toxic impact on lungs, including granulomas. The use of certain ingredients of animal origin (prohibited cattle materials) is prohibited.

There are substances whose use is not prohibited, but limited. Hexachlorophene is toxic and can penetrate the skin. Mercury compounds penetrate through the skin and accumulate in the body; they can cause allergic reactions and skin irritation or exhibit neurotoxic effects. The term "sunscreen" with sun protection substances (sunscreen in cosmetics) is a reason for these substances to be included in the section "Drugs" in order to apply to them the drugs regulation rules.

Just like the food industry, cosmetics uses a large number of different additives that perform the functions of emulsifiers, foamers, thickening agents, flavoring agents, antioxidants, disinfectants, preservatives, solvents, and so on, that is, the same functions that they perform as dietary supplements. Often, these supplements have the same notations as food additives, E***. In the case of their presence in cosmetic products, they exhibit the same properties. Also, some of them may be harmful or hazardous to health.

Some of them can cause allergic reactions, and others can make skin and hair too dry and cause itching and irritation. There can also occur more serious impacts. Some substances cause pathologic changes in the liver and kidneys. Also, substances with neurotoxic effects are known to exist. Of course, major attention should be drawn to those substances that are recognized as dangerous worldwide and whose harm is scientifically proven. But for some, mostly commercial, reasons they keep staying in usage. They constitute the whole group of carcinogenic substances and components that form carcinogenic substances entering into reactions with the cosmetic components of DEA, MEA, TEA, BHT, sodium lauryl sulfate, polypropylene glycol, and so on. Sodium lauryl sulfate is harmful because it can contain 1,4-dioxane (a powerful carcinogen directly related to dioxin—a very toxic chemical), as well as xenoestrogens and xenobiotics, that is, substances foreign to the body.

Cosmetic products always include colorants and fragrances. Often, these are synthetic substances. Dyes increase the attractiveness of cosmetics. They are labeled as FD & C or D & C, followed by color and number, for example, FD & C Red № 6. Many synthetic colorants are carcinogenic. Synthetic fragrances are designated simply as "flavor," but can contain up to 200 chemical ingredients. Some of these can cause headache, dizziness, rash, hyperpigmentation, coughing, vomiting, or skin irritation.

Even seemingly well-known and long-used Vaseline can cause alarm. Highly pure white Vaseline is safe, but in cosmetics the insufficiently treated yellow Vaseline may be used, which is carcinogenic. For consumers, it is very difficult to make the right choice. They have to rely on the honesty and reliability of the manufacturer.

Another danger of serious concern for experts is the lack of labeling and information concerning the presence of substances that may cause either endocrine diseases (endocrine disruptors) or asthma in personal care products (shampoos, creams, deodorants), as well as in household chemicals. In a study conducted by American scientists [30], a large number of nonlabeled chemicals were found in the household chemicals and in the articles of personal hygiene.

The scientists analyzed 213 products of 50 types to reveal if they contain parabens, phthalates, bisphenol-A, triclosan, ethanolamine, flavors, UV filters, glycol ethers, and phenols. In the examined samples, 55 pollutants were detected, which confirms a wide range of exposure to ordinary household chemicals and cosmetics. Highest concentrations of toxic substances were found in perfumes, air fresheners, and sun protection creams. Most of the chemicals in focus of the study were not included in the list of ingredients on the product labeling. This deprives consumers of the opportunity to make the right choice when shopping [31].

Interestingly, among the samples tested were so-called organic products. However, they also included the above-mentioned toxic chemicals that were not indicated on the label.

The situation is similar with goods sold in the Russian Federation [21].

5.1.7 Other Toxic Substances in Everyday Life

Apart from dangerous and harmful substances in the air, water, food, cosmetics, and detergents, one can also be faced with other toxic chemicals in everyday life.

Harmful substances that people may contact in everyday life and at work should be divided into two groups. The first group is the substances that can cause an adverse reaction of the organism, but their effect will terminate immediately, or shortly after, cessation of contact. Powerful cleansing and repair mechanisms of the human body either quickly, or slowly, depending on the particular substance and the condition of the body, remove the substance so that the contact with it can be forgotten.

The other group is the substances that produce irreversible changes in the body. There exist substances that irreversibly affect the liver and kidneys, nervous or circulatory systems, and so on. But in this book, we pay special attention to the substances that even in small doses can affect the genetic apparatus. These are mutagens, carcinogens, and teratogens. Their impact on a human is stochastic in nature. Just as in the case of radiation exposure (see Section 4.4), one can only be concerned of the probability of the effect. Whether these substances are going to manifest themselves during the remaining life of the individual is a matter of one's physical condition, one's state of the immune system, and the possible effect of other substances that enhance the action of the original carcinogen. It is, in the long run, a question of one's fate. Science is still unable to give clear predictions on this issue.

Within this book, it is impossible to give detailed toxicologic information on a large number of substances. But let's point out a few substances. We are going to focus on the special class of substances, on the simple substances, in fact, elements. There are elements that are definitely harmful; their danger is well known. These are some heavy metals such as lead, mercury, cadmium, and thallium. Some experts believe that they are harmful in any concentration, that is, there can be no maximum allowable concentration for them. Others only consider mercury as definitely toxic. Toxic properties of arsenic and antimony are also well known, but in certain circumstances, these elements (by their chemical definition, they are semimetals) are used as drugs.

5.1.8 Trace Elements

Let's turn to a group of so-called trace elements. In low concentrations, these elements are vital to the body, but greater concentrations can cause poisoning.

In the human body are found, in total, four basic elements (C, H, O, N), eight macro elements that are present in relatively large amounts (Ca, Cl, F, K, Mg, Na, P, S), and ten trace elements (Al, Co, Cr, Cu, Fe, I, Mn, Mo, Se, Zn). Basic and macro elements comprise 99% of the body weight. Trace elements are included, as a rule, in compounds, facilitating their involvement in metabolic processes [32].

Despite the vital importance of these trace elements, an overdose can appear dangerous. This, again, confirms the long-standing conclusion by Paracelsus: "There are no toxic substances; there are toxic doses."

For generality, it has to be added that, fundamentally, almost the entire periodic table is present within the human body. The chemical elements that are found with humans in very small amounts, in the range of $10^{-3}\%$–$10^{-5}\%$, are called trace elements, and when their concentration is less than $10^{-5}\%$ they are called ultramicroelements. The experts point out 81 items that can be found in the human body by careful analysis. Beside the above-listed, these experts mention the conditionally essential (vital, but harmful in certain doses) trace elements (Ag, Au, B, Br, Co, Ge, Li, Ni, Si, V)—10 elements in total. The remaining, present in nature elements from the periodic table, are referred to as conditionally toxic trace elements and as ultramicroelements. It is supposed that mercury (Hg) is harmful to humans in any amount; that's why it can be called a definitely toxic element [32].

In general, the effect of trace elements on the body is well illustrated by the graph in Figure 5.1.

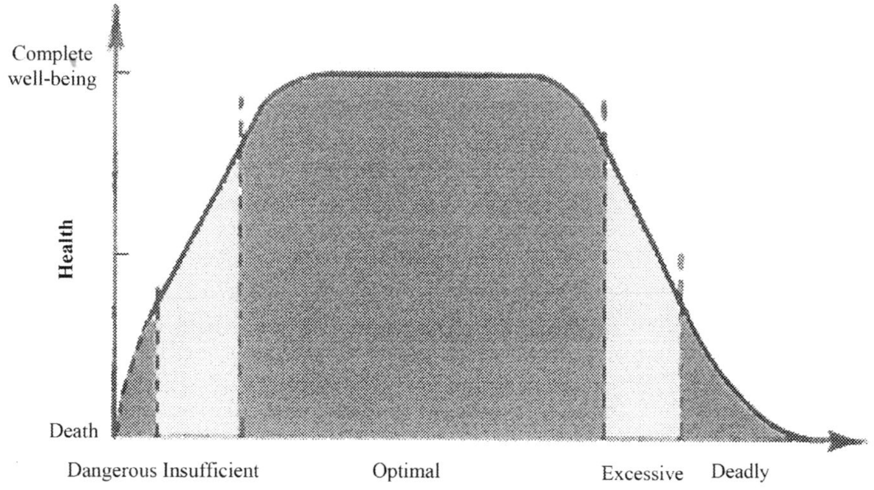

FIGURE 5.1 Biological action of some arbitrary trace element.

It should be noted that an absolute intake of trace elements is not the only factor causing the development of acute or chronic intoxication. "Safe" or "toxic" levels of each trace element essentially depend on certain conditions.

For example, if a trace element can accumulate in tissues, then the value for a minimum intake required for the occurrence of toxic effects is reduced. In the body, there are mechanisms that regulate uptake and excrete trace elements. Toxicity is also influenced by the duration of exposure, age, sex, nutritional status, immune status, physical activity, and the impact on the individual of other toxic substances (e.g., alcohol and tobacco).

Copper: Copper is part of important enzymes involved in metabolic reactions. Copper deficiency leads to anemia, poor bone state, and connective tissues. On the other hand, an intake of large quantities of copper (copper salts) can cause acute poisoning. Released to the gastrointestinal tract, a considerable amount of copper irritates the nerve endings in the stomach and intestines, causing vomiting, headache, and diarrhea. Chronic excess of copper leads to stunt, hemolysis, low content of hemoglobin in the blood, and tissue destruction in the liver, kidneys, and brain [33,34].

Zinc: As a part of many biologically active substances, is involved in several important processes that occur, for example, in the pancreas, where it stabilizes an insulin molecule or is involved in the transport of CO_2 by the blood and in releasing it into the lungs. During absorption of zinc, the formation of complexes with amino acids, peptides, occurs.

Chronic zinc poisoning is unknown. Random occasions of it are associated with the misuse of galvanized utensils for cooking acidic foods. It was pointed out that there is a great difference between the amounts of zinc entering the body and those capable of causing cumulative toxic effects. It was established that a zinc dose causing nausea (between 225 and 450 mg) is equivalent to zinc sulfate intake of about 1–2 g. The maximum allowed intake of zinc is 50–100 mg.

Cobalt: It is a component of the vitamin B_{12}. Deficit of B_{12} is most clearly manifested in the body in the areas of rapid cell division, for example, in bone marrow, hematopoietic tissues, as well as neuronal tissues, which can result in degeneration of the nerve fibers of the spinal cord and peripheral nerves. Deficiency of vitamin B_{12} affects normal formation of blood cells.

Toxic effects of excessive amounts of cobalt are rarely observed but can be seen particularly in cases of immoderate consumption of beer (beer foam was stabilized by cobalt salts) or when therapeutic doses of cobalt contained in drugs used to treat certain forms of anemia are exceeded.

Aluminum: The widespread occurrence of aluminum in the earth's crust, as well as its domestic use (kitchen utensils, equipment and packaging), is responsible for its presence in foods and beverages. Aluminum is found in almost all human organs and tissues. In moderate amounts, this trace element performs a number of important functions. Especially, it is involved in the building of bone and connective tissue and in the formation of epithelium.

For many years, aluminum was referred to as an element with minimal toxic effects, so the regulated level of daily consumption of this element was 500 mg. Over the years, however, based on observations and special studies that revealed the toxic properties of aluminum compounds, the recommended dose of daily consumption was reduced to

2–100 mg. According to the Expert Committee FAO/WHO, the weekly norm for the consumption of aluminum in food must not exceed 7 mg/kg.

For more detail on the toxic effects of aluminum, see Section 5.2.1.

Chromium: Chromium's characteristic property is its ability to manifest varying degrees of oxidation in compounds. Most widespread are two forms: Cr(III) and Cr(VI). In food products, the element is usually present in the form of Cr(III). The effects on the body of the two most widespread oxidation states of chromium are different. If Cr(III) is an essential element, Cr(VI) is toxic. The main role of Cr(III) is to maintain a normal level of glucose in the body. Deficiency of Cr(III) leads to disruption of glucose and lipid metabolism and can cause diabetes and atherosclerosis. Easily penetrating into the cell membranes, Cr(VI) at prolonged exposure (which often happens in industrial environment) can cause lung cancer and performs mutagenic and teratogenic effects. This is attributed to the formation of complexes of chromium with nucleic acids, proteins, and so on, thus causing damage to DNA.

Selenium: Selenium's main property is its antioxidant capacity, that is, an ability to remove free radicals from the body and to resist oxidation processes. A powerful antioxidant effect of selenium is associated with the fact that it is the main element of the antioxidant enzyme MPO (glutathione peroxidase), which performs high activity in body tissues. The lack of it causes weakness, fatigue, and dizziness. Selenium is most effective in conjunction with vitamins E, C, and other antioxidants. Intoxication with excess amount of selenium is manifested by loss of hair, changes in nails, skin (dermatitis, eczema, burns, and allergies), tooth decay, and negative effects on the nervous system.

Also, it has to be noted that the impacts of various trace elements on the body can be interrelated. For example, the antagonism between zinc and copper has been noted, when overconsumption of zinc affects the balance of copper, which, in its turn, affects the performance of cholesterol in blood plasma, as well as the activity of enzymes that contain copper. Copper consumption should be associated not only with the content of zinc but also with that of molybdenum, which also affects the metabolism of copper.

5.2 The Toxic Effects of Chemicals

In Section 5.1, it was shown what a vast number of different substances a human being is surrounded with in nature, at home and at work—some of which are toxic. Information on these substances, their chemical and physical properties, and their toxicity can be found in the manuals on toxicology [35] and in databases of toxic substances [6,7]. The number of absolutely poisonous substances, as was already mentioned, is extremely low. Most of the substances that we call poisonous can, in certain doses, be harmless, useful, or even necessary for normal development and functioning of the body.

5.2.1 Metals

5.2.1.1 Arsenic

Let's start with arsenic. Authors of the book *Poisons – Yesterday and Today: Essays on the History of Poisons* [36] devote a separate chapter to arsenic, titled "Poison of Poisons – Arsenic."

Arsenic is a chemical element of group V of the periodic table, a chemical analog of nitrogen, phosphorus, and antimony. Some arsenic compounds are extremely toxic, such as arsenic oxide, As_2O_3, its sulfur compounds, and gaseous arsenic hydride AsH_3. Other compounds are completely harmless. Pure arsenic is also relatively inert. It is believed that in small doses arsenic is useful because in the body it performs a number of important functions.

The typical level of arsenic in the air is 1–3 ng/m^3. In urban air, the level may get higher, up to 20–100 ng/m^3. In drinking water, it is ~2 μg/L. The consumption of arsenic with water, with breathing, or with food, can lead to a situation when a human being who was not actually exposed to arsenic can get <1 μg/L of it in blood, <100 μg/L of it in urine, <1 ppm of it in nails, and <1 ppm of it in hair.

According to various sources, the daily requirement for arsenic is 50–100 μg. The toxicity threshold, which is the minimum of dangerous amount of arsenic (minimal risk levels [MRL]), is at about 5 mg/day. For the organic compounds of arsenic, the MRL comprise fractions of milligrams or even micrograms, depending on the compound, the duration, and the way by which arsenic enters into the body. A dose of 50 mg or higher can cause death.

No other chemical element in the whole history of mankind has, perhaps, such a sinister reputation. Arsenic is traditionally associated with poison. A rather great attention was paid to it by men of letters. As it is pointed out in the article "Arsenic and Human Health" [37], the number of victims of this chemical in literary works may exceed the number of those who ever actually died from it. If a character of an Agatha Christie's novel died due to poisoning, it usually had to be from arsenic.

From the large number of detective stories relating to arsenic poisoning, we'd better choose the story of a possible poisoning of Napoléon Bonaparte, since in this case, the investigations were pursued at quite a decent scientific level [38]. The emperor died in 1821 on the island of St. Helena at the age of 51. The officially announced cause of death was stomach cancer. But even then it raised serious doubts. In the early 1960s, a radiochemical method of the neutron activation analysis (NAA) was applied to the analysis of the elemental composition of one of the hairs from Napoléon's head [39]. Studies have shown that in his hair the level of arsenic was significantly higher than normal (10.38 ppm). On this basis, the suspicion arose that the real cause of death was arsenic poisoning.

In the 2000s, a new study of Napoléon's hair was carried on, now with the help of a modern version of NAA. For the tests, two hairs were taken from the strand of his hair cut the day after the death. The hairs weighed, respectively, 0.1722 and 0.2151 mg. The concentrations of five elements in them, arsenic, mercury, chromium, antimony, and zinc, were detected simultaneously. The concentration of arsenic found in these two hairs was 1.85 ± 0.11 and 3.05 ± 18 ppm, which exceeds the normal level for the local population. The measurements even made it possible to detect unequal distribution of arsenic inside one hair. On the assumption that human hair grows with an average rate of approximately 0.4 mm/day, the time of poisoning can be, actually, identified.

However, the arsenic concentrations in the second series of measurements were not so high, and yet another version has undergone a more detailed analysis, this time it was the poisoning by compounds of antimony and mercury, which also have a toxic effect.

About 2 months before his death, Napoléon was taking potassium tartrate of antimony, which he was given to restrain vomiting. Besides, the doctors were giving him calomel and orgeat to combat his constipation and thirst. Constipation and thirst are symptoms of arsenic poisoning. Calomel is a mercuric chloride Hg_2Cl_2. Orgeat is a drink that contains bitter almond oil. In the stomach, this oil is hydrolyzed into cyanide (hydrocyanic) acid. When calomel and orgeat mix in the stomach, mercury cyanide is formed. So, either the hydrocyanic acid or mercury cyanide could be the main cause of Napoléon's death. The toxic effect of cyanide is based on the fact that, as a result of chemical reactions between cyanide and cytochrome oxidase, the final link of oxidation gets blocked and tissue hypoxia develops. Oxygen is delivered by arterial blood to the tissues in sufficient quantities, but it does not get absorbed and passes unchanged into the vein course. For a human being, the minimum registered lethal dose of hydrocyanic acid is ~1 mg/kg and that of potassium cyanide is ~3.5 mg/kg.

The analysis showed that the mercury content in hair (3.98 ± 0.29 ppm in one of the hairs) was within the normal range (0.5–10 ppm). Antimony concentrations in both hairs (4.47 ± 0.27 and 4.32 ± 0.38 ppm) though were relatively high. It is considered that the normal level of antimony content in hair for the local population is between 0.016 and 1.3 ppm.

Most recently, new data on the content of arsenic appeared. Physicists have compared different strands of Napoléon's hair (cut in his childhood in Corsica, at his mature age while being in exile on the island of Elba, and the next day after his death on St. Helena), the hair of his son, his wife, the Empress Josephine, and of 10 different people living in our days. As it turned out, the hairs of Napoléon and those of his contemporaries contain one hundred times more arsenic than the hairs of our contemporaries. The quantity of arsenic in the strands of the emperor, cut when he was a child, does not differ from the amount that was found in the postmortem sample of his hair. It is not yet entirely clear how arsenic got into the hair, from inside his body or from outside. Obviously, the family of Napoléon often in their everyday lives somehow came into contact with considerable quantities of this poison. Highly toxic arsenic compounds were used then for painting wine barrels and ceramic tableware, and were added into wallpaper paste and powder.

So, the arsenic hypothesis of the emperor's death at this stage of research apparently disappears. The version with antiemetic overdose, possibly unintentional, remains in force.

The above example was meant to draw the readers' attention to the possibilities of modern methods of analysis, to the small size of the studied object combined with the high sensitivity and accuracy of the research method.

5.2.1.2 Lead

Toxicity of lead is well known. There even exists an opinion that it was lead that promoted the decline and degeneration of the Roman Empire: researchers think that the Romans kept being poisoned from generation to generation both by water flowing through lead coverings in aqueducts and by wine stored in leaded vessels.

Nowadays, a great harm to the environment and human health comes with the use of tetra-alkyl derivatives of lead as antiknock additives in gasoline. They are now prohibited for use but, nonetheless, the growth of lead production continues. Lead in large amounts is used for protective screens in nuclear physics. Having a high atomic number and a high density, lead absorbs gamma radiation very well.

Basically, it is considered that lead is harmful in any concentrations, particularly for children. Nevertheless, some amount of lead is present in the blood of a normal healthy human being. Average lead levels in an adult's blood is $<20\ \mu g/100\ mL$; with children, it is less than $10\ \mu g/100\ mL$.

Up to now, some companies in different countries continue to produce and sell the paints that contain dangerous levels of lead. Widely known is the white lead; usually, it is a mixture of carbon monoxide and lead oxide. Lead abatement campaign is supported by the U.S. Environmental Protection Agency [40]. Starting from 2009, the lead content in paints in the United States is legally restricted to 90 ppm. In some samples tested in 12 countries, lead level exceeds 10,000 ppm. According to modern research, exposure to even very low levels of lead leads to negative consequences for children's health, and a safe level of lead content is a nonsense.

Lead can also be found in lipstick. Recent studies in the United States showed that 61% of lipsticks from well-known manufacturers contain high levels of lead [41].

5.2.1.3 Mercury

Mercury is a fairly common element that is found in air, water, and soil. It occurs either as a free metal mercury, or in the form of both inorganic and organic compounds. One of the main sources of mercury in the air is burning of coals that contain mercury. In the United States, some 40% of mercury emissions are due to this. The rest occasions happen because of mercury spills and bad operation and disposal of mercury-containing devices (mercury thermometers, barometers, switches, fluorescent lamps). Mercury from the air gets directly into the water or is washed into the water from the soil and, under the action of microorganisms, is converted into methyl mercury, which accumulates in fish and other water organisms. Consumption of fish and shellfish is one of the main ways of mercury intake. Another way is the inhalation of mercury vapors, but this is only possible as a result of an accident.

Mercury has been used in medicine for centuries, in order to make amalgam in manufacturing of mirrors and in some other technological processes. Silver amalgam is widely used in dentistry as a dental filling material and is still used, although a lot less frequently.

An exposure to mercury can be, in particular, illustrated by the following example. In the 1800s there appeared the expression "Mad as a Hatter," which owed its origin to the fact that the master-hatters—who processed felt hats with the help of mercury—often had mental disorders. The main targets of mercury are the brain and the nervous system, since mercury is a neurotoxin. It is believed that mercury is more toxic to humans than lead or arsenic.

The maximum content of mercury in drinking water permitted by the U.S. Environmental Protection Agency is 2 parts per 1 billion. Usually in developed countries, the tap water fulfills this condition, but if water flows from wells, the limit sometimes can be exceeded. For example, the drinking water studies in the state of New Jersey in the United States showed that in 265 of 2239 wells in use, the mercury levels are higher than normal.

5.2.1.4 Cadmium

The studies performed in the United States indicate that people in general consume with food approximately 13–16 mg of cadmium per day. The individuals who are not exposed to cadmium have cadmium traces in the urine of 0.59–0.77 μg/L, in the blood 0.09–0.11 μg/100 mL, in hair 0.83–1.10 μg/g, in kidneys 21 μg/g, in liver 1.2 μg/g, in muscles 0.067 μg/g, in the pancreas 0.58 μg/g, and in fat 0.040 μg/g. Cadmium concentration in the hair is higher in men than in women. It increases with age and is higher with smokers than with nonsmokers.

High doses of cadmium are hazardous for health and even for life. It can enter the body of those who work with a lot of cadmium dust or fumes, with heating of products containing cadmium or having a surface coated with cadmium, with welding or soldering metal parts containing cadmium, or with cutting these. Cadmium is a lot more dangerous when inhaled than when ingested.

Symptoms of poisoning appear within 1–10 hours, and the difficulty with breathing can lead to death. If one manages to avoid death, the main symptoms slowly disappear within a week.

The prolonged exposure to small doses leads to kidney damage and to the increased risk of lung and prostate cancer.

It is known that cadmium has a high probability of absorption of neutrons and therefore is used as a shielding material in nuclear physics. Physicists working with neutron instrumentation can come in contact with cadmium.

5.2.1.5 Thallium

Thallium is used in manufacturing of electronic devices, as well as in production of glass and also for certain medical procedures. Thallium can be released into the environment from burning coal and smelting of metals, which may contain traces of thallium. It can stay in the air, water, and soil for a long time. Eventually, thallium gets into plants and furthermore into the food chain. Inhalation of thallium from the air and consuming it with food are the main ways for thallium to get into the body.

Thallium is the main activator for alkali halide scintillators, for example, NaI(Tl). A TlI impurity with a concentration of ~0.1% is introduced into a bath consisting of NaI salts and then a crystal grows. The physicists working with such scintillators may be subject to thallium exposure. All alkali-halide crystals look similar. They are large, transparent, single crystals. If in a lab they happen to be mixed up, one has to determine the type of crystal by very labor-intensive instrumental methods. The author once witnessed how, in one laboratory, this procedure was simplified by identifying the type of crystal by tasting it. Some unique individuals could, by licking a crystal, roughly determine the thallium concentration. Thus, the ways of thallium into the body are varied and inscrutable.

Exposure to large doses of thallium has a certain effect on human health. Studies on the employees operating with thallium in the industrial process, showed particular effects of thallium on the nervous system, lungs, heart, liver, and kidneys. Too high doses cause death. The effects of prolonged exposure of thallium in small doses are unknown. The relevant American organizations (Occupational Safety and Health Administration [OSHA] and the American Conference of Governmental Industrial Hygienists [ACGIH]) set up the limit for thallium quota in the air of workplaces as 0.1 mg/m^3.

5.2.1.6 Aluminum

Aluminum is a chemical element of Group III of the Periodic System: with atomic number 13, it is a silvery-white light, ductile metal. It is chemically active; in air it gets covered with a protective oxide film. In terms of its natural occurrence, it occupies the fourth place among the chemical elements and the first among metals (8.8% of the mass of the Earth's crust). Several hundred minerals of aluminum (aluminum silicates, bauxites, alunites, etc.) are known.

Aluminum is widely used in everyday life (utensils), in engineering, in the food industry (foil), and in water disinfection. It also has a lot of other applications. According to the experts, the level of consumption of aluminum is heading the list among the most important metals. The food supplement E173 is aluminum. In the food industry, the aluminum as the additive E173 may be used as a colorant for the decoration of confectionery, dragées, and cakes. The dye E173 imparts to the product a shiny silver color.

Although aluminum is not included in the group of recognized toxicants (heavy metals), in large doses it exhibits significant toxicity and if it penetrates into the brain, even small quantities of it are enough to cause tangible harm. Aluminum's neurotoxicity was discovered more than 100 years ago. Lately, aluminum was credited with initiating several more pathologies. Thus, there are suggestions of a possible connection between aluminum and Alzheimer's disease; the hair of exposed patients reveals an increased content of aluminum. However, despite the current extensive research, a comprehensive answer is still inaccessible. Nevertheless, it is obvious that aluminum is neurotoxic even in the smallest quantities, for it has a long history of well-documented harmful effects. According to the American Academy of Paediatrics, "it is now considered that aluminum interferes with a variety of cellular and metabolic processes in the nervous system and other tissues" [42]. This has caused some researchers to suggest an existence of a possible link between aluminum and autism [43].

Some vaccines administered to infants are rich with aluminum. The vaccine of hepatitis B is injected at an infant's birth, then at 2 and 6 months of age, and each dose contains 250 μg of aluminum. A combined vaccine against diphtheria, tetanus, and pertussis is injected to the babies at 2, 4, 6, and 15 months, and each dose contains 625 μg of aluminum, and so on. As a result, the infants during the first 18 months of their life receive about 5000 μg (5 mg!) of aluminum. Such high aluminum content can stimulate both autism and some other neuropathology.

In the mid-twentieth century, there were suggestions of a relation between aluminum and acceleration—the increase in growth and early maturation of children. The researchers have paid particular attention to the fact that due to the omnipresence of aluminum cookware, children just have to lick the aluminum along with the milk.

The studies of toxic properties of aluminum and its metabolism are continuously increasing in number. If in the 1970 and 1980s one could find 140 publications on "aluminum toxicity," in the 1990 and 2000s there are already 1035 of them. Then in October 2012, a Google search offered 318,000 sites for a search request of "aluminum toxicity."

It has been found that aluminum can replace some metals in enzymes and that it can provoke some diseases of the lungs and eyes, eczemas, and dermatitis. When the concentration of aluminum in the blood serum is equal to 40–45 μg/L, the kidney tissue can be damaged, and at 150–200 μg/L, the development of brain lesions is possible. Aluminum inhibits absorption of some vitamins and minerals as well as of some sulfur-containing amino acids. The most dangerous is the effect of aluminum on newborn babies and children with impaired renal function, as well as on adults with diseases of the central nervous system and renal insufficiency.

Especially important today is the problem of the remote effect of substances that had been considered harmless (benign) until recently. These effects can be manifested both in the form of selective damage to an organ and as a general disruption of functions of the whole organism.

It is believed that the usual aluminum intake of a human is from 3 to 10 mg/day. The main source of aluminum is tap water. The daily requirement of an adult is 30–50 μg. A toxic dose is higher than 50 mg.

5.2.2 Toxic Substances

Apart from toxic elements, nature has created a lot of poisonous compounds. Within this book, it is impossible to give an overview of most of them, let alone all. The interested reader is referred to the databases of toxic materials [6,7] and to toxicology guidelines [35].

However, some groups of obvious poisons have to be mentioned and some information on the mechanisms of toxic action should be considered.

5.2.2.1 Chemical Weapons

It is well known that chemical weapons are prohibited for the use throughout the world but huge stocks of these weapons have been accumulated and huge amounts of money have to be spent for their destruction. Sometimes accidents happen at warehouses of chemical weapons or at the plants that produce them. Then poisonous gases come into the atmosphere, causing substantial damage.

Besides, up till now the police and special units of internal forces in many countries use tear or nerve gases in order to deal with protesters.

Chemical weapons include substances with different ways of exposure: lethal poisons, temporarily incapacitating toxic substances, and psychochemical substances.

Chemical weapons that cause death are especially dangerous. These are, primarily, poisons with nerve and general-toxic action. At high concentrations of toxic substances or prolonged exposure to an infected site, toxins start to destroy the human body from the inside, causing its systemic poisoning. The neuroparalytic agents sarin, soman, V-gases, phosphagen, and tabun are regarded to the deadly poisons. Each of these toxins exhibits its specific effect and possesses specific chemical characteristics. The deadly chemical weapons used in the Vietnam War by U.S. troops were phytotoxicants.

The blister agents and the psychochemical toxins cause temporary paralysis of the enemy. Iprite performs the most powerful effect on the human body, causing skin lesions. The toxic agent iprite is a liquid but, depending on the temperature, it may start to evaporate, which leads to saturation of the room or area with poisonous vapors. Iprite is virtually colorless, and in a pure state its smell is quite imperceptible. But the chemical iprite has an apparent smell of mustard, which one gets used to rather quickly. Iprite has a slow effect on the body. It primarily affects the mucosa of the eyes, causing discomfort and tears. Getting inside by inhalation, iprite affects internal organs, causing dizziness, severe pain in the heart and general malaise. Iprite has a particularly strong effect on the skin of humans. Also belonging to poisons is the toxin BZ that gets in through the skin and causes a variety of allergic reactions. This chemical causes skin lesions and can also cause hallucinations.

The most common drug among psychochemical substances is LSD-25. Toxins of this group have a temporary effect on the human body. Various psychochemical poisons can cause different reactions, and they are able to inhibit the will and to affect the brain. There are algogenic drugs that excite severe pain. Anxiogenic drugs can force the enemy to flee in panic from the battle place. Aphrodisiacs that promote human sexual desire can bring people out of balance. Anesthetic agents quickly immerse the enemy into a real anesthesia. Some psychochemical substances can completely immobilize people by relaxing their muscles, or by disrupting their memory. Neuroleptics contribute to inhibition of action and thought.

In 1993 Russia signed and then, in 1997, ratified the Chemical Weapons Convention. In this regard, it adopted a program of destroying stockpiles of chemical weapons built up over many years. Initially, the program was designed for the period till 2009. However, because of underfunding, the program was changed. According to press reports, by mid-August 2012, more than 26,000 tons of warfare chemical agents were destroyed, which was more than 65% of all the chemical weapons that Russia inherited from the USSR.

5.2.2.2 Poisons

Amid the strongest poisons are cyanides. The most well-known toxic substance of this group is considered to be formonitrile, hydrocyanic acid [HCN]. Hydrocyanic acid mainly affects human blood. Poison inhabits the ability of blood to carry oxygen, and human organs just cease to function because of the lack of oxygen. Apart from hydrocyanic acid, the commonly used substances of this group are actual cyanides (cyanide salts), cyanides of sodium (NaCN), and cyanides of potassium (KCN).

Potassium cyanide was popular among scouts and spies, as just one capsule of it caused instant death of a captured agent, and helped to avoid torture. According to the detective stories, a capsule was usually sewn into a shirt collar so that the agent could reach it with his teeth being tied hand and foot. The median lethal dose (LD_{50}) of potassium cyanide is 1.7 mg/kg. It is well known that the people who planned to kill Gregory Rasputin tried to poison him with several lethal doses of cyanide, but which did not kill him. Therefore, he was drowned by his antagonists.

Strychnine ($C_{21}H_{22}N_2O_2$) is a poison well known from literature. For example, the heroes of one of the stories by Jack London poison each other with strychnine. P. G. Wodehouse titled one of his novels *Strychnine in the Soup*. Strychnine blocks glycine sections of *N*-methyl-D-aspartate (NMDA), the receptors on the membranes of neurons in the spinal cord and brain, thereby suppressing the regulation of inhibitory processes in the central nervous system. The median lethal dose (LD_{50}) is about 1 mg/kg. Strychnine, being a strong poison, in small doses can have a very beneficial effect on the human body and is used as a medicine. Strychnine was chosen by Agatha Christie as a favorite means of murder in her stories. These days, strychnine as a criminal poison can be found only in fiction.

To estimate the order of lethal doses of poisons, here's the list of specific figures: potassium cyanide, 120 mg; cobra venom, 14 mg; carpet viper's poison, 5 mg; palytoxin, $\sim 1\ \mu g$.

5.2.3 Mechanism of Toxic Action

Once ingested, toxic substances enter physicochemical reactions, usually it is dissolving in an aqueous or lipid media of cells and tissues or a chemical interaction with certain structural elements of the organism, called "receptor" or "target." One can see that the concept of a "target" also appears here as in the case of radiation damage. However, the radiation here gets to a target directly through any part of the body and the chemical toxicants are delivered to a target by one of the ways of substance transfer within a living organism.

A number of chemicals act as poisons of local action, destroying tissues through direct contact. Another category is the poisons with systemic effects: they enter the bloodstream and affect the heart, kidneys, nervous system, or other vital organs. This type includes cyanides.

5.3 Methods of Carcinogen Screening

Since chemical carcinogenesis comprises a significant, or perhaps the major, part of all cancers, it is important to determine what substances have carcinogenic properties, what is the activity level of these substances, where they come from, how one can measure the concentration of each of the suspected substances, and how one can eliminate or drastically reduce human contacts with these substances. Also, what arrangements are needed to prevent a disease?

The above-mentioned matters highlight the two problems: first, define the carcinogenic risks of specific substances and, second, how to identify substances with already-known carcinogenic risk in the human environment. To solve the second problem, the full power of physicochemical analysis methods is used. To solve the first problem, the different ways to determine the carcinogenicity of chemicals are instrumental. The experts call these a screening for carcinogens.

The known methods of screening are as follows:

1. Epidemiological studies.
2. Long-term experiments on animals.
3. Physicochemical methods.
4. Short-term tests (STTs) in vivo and in vitro; bacterial tests.
5. Correlation of the molecule structure and its biological activity (quantitative structure–activity relationship [QSAR]).

5.3.1 Epidemiological Method

Experiments on living people are impossible. However, life itself unwittingly carries out a huge experiment, offering a human being various living conditions. Depending on where they live and their lifestyle and professional activity, people are exposed to a variety of factors contributing to the emergence and development of disease. The analysis of these factors is possible by collecting statistical information on special features of a disease distribution in connection with the distribution of possible damaging factors. This is called the epidemiological method.

The basic principle of epidemiological studies is to compare the frequency of a particular disease among people exposed to a damaging factor and among those who did not contact this factor. If the frequencies are similar, it gives reason to believe that the analyzed factor is not the cause of the disease. That does not mean that this factor is safe always and for all, but at the given time and for the given population group, there is probably no reason to get scared of it. If the frequency of disease in the group of people exposed to the factor is significantly higher than in the unexposed group, it means that this factor may be the cause of the disease. If the frequency of the disease in the studied group is lower than in the control group, then it can be accepted that the factor has a protective effect (see Section 3.1).

The meaning of "higher," "significantly higher," and "lower" and the similar other in this context is defined by the rules of statistics (see Section 3.1).

Epidemiology started and established itself on the basis of the struggle against infectious diseases such as smallpox, plague, and cholera. Importantly, epidemiology is not only the study of cancer-causing factors but also the development of methods to combat them. Currently, epidemiology has been used successfully in the analysis of radiation damage, of exposure to chemical carcinogens, diabetes, hepatitis, malaria, cardiovascular diseases, and so on.

On the one hand, it can be expected that by this method the most reliable information is obtained, since it is a human himself that gets exposed to the effects of a potential carcinogen, and the final result of this impact is exactly the effect that is the purpose of the research, that is, the induction of malignant tumors. That's why experts believe that any results of experimental studies, of experiments on animals, bacteria, and other systems should be confirmed by epidemiological studies. On the other hand, because the effect is unintended, it is often difficult to determine the dose, duration of exposure, confounding factors, and many other aspects required to be taken into account by the competent studies.

Epidemiological studies of radiation effects on humans are discussed in Section 6.5.3. This section focuses on the epidemiology of chemical carcinogenesis.

Since the impact of chemicals on the human body is not to be set intentionally for research purposes, we have to work with what we've got, and thus, with those carcinogens that are present in the environment and/or in the workplace. Unfortunately, there are quite a lot of chemical factors in the environment that individuals have to deal with, sometimes regardless of their wish and sometimes because of their ignorance, negligence, and often stupidity. Epidemiologists and specialists have enough to work with. In epidemiological studies, it becomes possible to collect lots of data that are useful to science and to the fight for the healthy population.

All carcinogens that are to be analyzed in the framework of the epidemiological method are divided into two groups for convenience: professional and environmental. Nevertheless, it is not always possible to clearly identify the differences between the two groups. Some professional carcinogens are also present in the surrounding environment. However, the level of exposure at the workplace is substantially higher as usual. It is necessary to consider the role of climatic, geographical, and geochemical factors in the spread of malignant diseases.

5.3.1.1 Cancer and Occupational Activity

Perhaps the best known is the association of cancer hazard with occupational activity. Actually, with the establishment of such a relationship started the history of epidemiological studies in this area. At the end of the eighteenth century, the English physician Percival Pott detected a connection between scrotum cancer in chimney sweepers and their job. The fact is that in Britain, the houses were heated by coal. And because chimney cleaning was considered difficult for the adults, the chimney sweepers delegated their job to the adolescents. Thus, young chimney sweeps got soiled in coal soot and, as a result, they got skin cancer at the relatively young age of 20–25 years. Most often, cancerous lesions appeared on the skin of the scrotum, where sweat with dissolved carcinogens accumulated. As a result of these observations in England, child labor in chimney sweeping was banned. After years of fighting for the abolition of child labor and for the use of more advanced working methods, cancer cases among the British chimney sweepers was graded up to that with other segments of the population.

Pott's considerations were supported by Yamagiwa and Ichikawa, who induced skin cancer in animals with the use of coal tar. Later, Cook and Kennaway selected from the same tar the active agent, carcinogenic polycyclic hydrocarbons. Benzo(a)pyrene isolated from coal tar became a classic carcinogen [44].

By monitoring diseases in people of various professions, cancer of the ears was diagnosed in the unskilled workers who carried bags of coal and bladder cancer in workers associated with the production of aniline dyes.

The professional cancer also includes the so-called winemaker's cancer. It is caused by arsenic that is used to combat the grape aphid phylloxera.

The historians of cancer point out that, back in 1770, Italian doctor Bernardo Ramazzini linked breast cancer in the nuns with celibacy and unrealized fertility. Then, in 1842 Rigoni–Stern obtained quantitative estimates of the risk of uterine cancer in the nuns and confirmed that uterine cancer was significantly more prevalent in the nuns than in other women in Verona. These cases can be attributed to both professional cancer and lifestyle.

In 1888, Hutchinson reported cases of skin cancer among patients who were treated with drugs that contained arsenic.

As early as in the middle of the sixteenth century, the "mountain sickness" of miners was described, and 300 years later, in the nineteenth century, Harting and Hesse found out that this disease is a lung cancer and is associated with the work in conditions of high radioactivity.

Eventually, elevated cancer rates have been found among a number of people in occupations engaged in the processing of certain minerals (nickel, chromium, beryllium, cadmium, cobalt, arsenic, etc.). It was experimentally shown that these substances exhibit carcinogenic properties.

All these observations have led to the introduction of the concept of "occupational cancer."

Before presenting information on the carcinogenicity of any substance, let's point out that in the classification of carcinogenicity, all substances are subdivided into five groups:

- Group 1 includes the substances, mixtures, and factors proven to be carcinogenic for humans.
- Group 2A includes substances and factors that are "very probably carcinogenic for humans."
- Group 2B comprises those that are "probably carcinogenic for humans."
- Group 3 includes those that are "not classifiable as to their carcinogenicity for humans."
- Group 4 represents those that are "not carcinogenic for humans."

Note the words "for humans." This means that experiments on animals, bacteria, and other ways of evaluation of carcinogenicity, which are discussed in Sections 5.3.2–5.3.5, are not the final verdict on the substance. For more information about the regulation of carcinogenic security, see Section 9.2.

The studies of occupational cancer revealed a few obvious problems. Obtaining reliable results was hindered by population mobility and difficulties tracking down all possible movements. In epidemiological studies, it was difficult to determine the dose received by the individual and the duration of exposure. Perhaps the biggest problem is the inability to take into account all the additional factors that sometimes may play a critical role;

they are lifestyle, smoking, drinking, culinary preferences, and other possible carcinogens in air, water, and food that are not related to professional activity. Careful analysis of possible side effects and complicated mathematical processing of results enable the epidemiological method to produce very valuable and very important results [45,46].

However, until recently, the experts, while identifying the active substance that determines the carcinogenic hazard of a certain industry, have been forced to express themselves in the subjunctive sense or by some other way to show their lack of confidence in the correctness of the statements. For example, carcinogenicity in the rubber industry is, *probably*, due to the use of 2-naphthylamine. Carcinogenicity in the coal-to-coke conversion, in aluminum production, and in the foundry industry is, *apparently*, determined to be the influence of PAH on workers. The workers in the shoe industry are at increased risk of leukemia and lymphoma, which is most likely related to exposure to benzene [46].

The workers of the wood industry are facing a significantly increased risk of nose and sinus cancer. There are no data on specific carcinogens affecting those who work in such industries. Most likely, the dust resulting from the processing of wood has an irritating effect on mucous membranes. D. G. Zaridze reports on the research conducted in his laboratory of epidemiology and prevention at the N.N. Blokhin Russian Cancer Research Center. A study on the workers of a Moscow shoe factory indicates an increase in morbidity and mortality from liver cancer, which is likely related to exposure to chloroprene (CP). Similar studies were conducted at the Yerevan plant, "Nairit," producing CP. Statistical analysis confirmed the existence of a cause–effect relation between exposure to CP and liver cancer [46].

Epidemiological study of morbidity and mortality in workers of the printing industry showed a significant increase in the risk of death from pancreatic cancer in male typesetters that were exposed to dust and vapors of inorganic lead. An increase in mortality from cancer of stomach and bladder and skin melanoma was revealed in female print workers exposed to paint-oil sprays and paper dust containing PAHs. Among workers of bookbinding workshops that were exposed to adhesive vapors and paper dust, there was a statistically significant increase in risk of death from esophageal cancer and ovarian cancer.

5.3.1.2 Lifestyle

Let's start the analysis of the impact of lifestyle and personal characteristics on carcinogenesis with a well-known event. Much has been written and spoken about it. And it has lots of staunch adversaries. However, the author believes it wouldn't go amiss to recall this event and to offer some, perhaps not so widely known, information about the dangers of it. It is smoking.

5.3.1.2.1 SMOKING

It was believed that the major harmful components in tobacco are nicotine and tar. For a long time, the content of nicotine and tar in tobacco products was not at all regulated. In 1989 the Sanitary and Epidemiological Inspectorate of the USSR (Sanepidnadzor) adopted some recommendations in order to limit the level of concentration of tar and nicotine in cigarettes: 19 mg of tar per cigarette and 13 mg of nicotine [47]. Today the standards of GOST 3935-2000 GN 2.3.2.1377-2003 are in force: not more than 0.9 mg of nicotine and 10 mg of tar per cigarette are allowed. The similar standards, 1 mg of nicotine and 10.1 mg of tar, exist in Europe. Control over the quality of cigarettes and over the chemical composition of tobacco smoke lead to the decrease of morbidity and mortality from lung cancer in Russia. Zaridze points out that since the number of smokers in Russia did not reduce, quality control is the only reason why the mortality rate might have fallen. But further reducing the concentration of these substances is meaningless. Tobacco contains carcinogens that are associated neither with nicotine nor tar.

Apart from the nicotine, the composition of tobacco smoke includes several dozens of toxic and carcinogenic substances: polycyclic aromatic hydrocarbons (PAHs) such as benzo(a)pyrene, dibenzanthracene, benzofluoranthene, dibenzo(a)pyrene, aromatic amines (naphthalenamine, aminobiphenyl), volatile nitroso compounds, tobacco-specific nitrosamines (*N*-nitrosodimethylamine, *N*-nitrosodiethylamine, *N*-nitrosopyrrolidine), vinyl chloride, benzene, hydrazine, urethane, aldehydes (formaldehyde, acetic aldehyde, crotonaldehyde), phenols, chromium, nickel, cadmium, polonium-210, uranium-235, uranium-238, arsenic, free radicals, and so on. Some of these substances are contained in tobacco leaf; others are generated by processing and combustion. Tobacco smoke is a mixture comprising a few dozen of toxic, mutagenic, and carcinogenic compounds that cause a variety of effects in different organs. This explains the universality of tobacco intoxication and the fact that it is associated with the development of more than 40 different diseases and disorders in humans.

The temperature of tobacco burning in cigarettes is very high when inhaling and is significantly lower between the puffs. This determines the different concentrations of chemicals in the main and the side smoke streams. The side stream contains more nicotine, benzene, and PAHs than the main stream. Most of the carcinogenic and mutagenic substances are contained in the solid phase of tobacco smoke.

The difference in the chemical composition of main and side streams of smoke explains the particular harmfulness of tobacco smoke for the nonsmokers who occur in the same room with the smokers. This fact is confirmed by extensive epidemiologic studies both in Russia and abroad.

So epidemiological research on the dependence of health on lifestyle has reliably shown the association between cancer and smoking. Risk of cancer development in smokers compared with that in nonsmokers is many times higher. Various studies offer the numerical values from none up to 16 times. The preferential localization of tumors in smokers is in the lungs, although other organs may be also affected.

5.3.1.2.2 ALCOHOL

The correlation of mortality from malignant tumors with the consumption of alcohol per head is also considered established. Excessive consumption of strong alcoholic beverages increases the risk of cancers of the oral cavity, pharynx, larynx, esophagus, liver, pancreas, and stomach. There are suggestions that alcohol increases the risk of breast cancer. There is a synergism between the carcinogenic effect of consumption of alcoholic beverages and smoking. On the basis of existing experimental and epidemiological data, the working group of the International Agency for Research on Cancer (IARC) came to the conclusion that the consumption of alcoholic beverages can be referred to the Group 1 of proven carcinogens.

The alcohol hazard, in particular, is proved by the fact that the mortality from cancer among members of religious groups who abstain from smoking and consumption of alcoholic beverages is significantly lower than among the general population.

It is also significant that ethanol itself is not carcinogenic. However, it plays the role of a promoter for carcinogenesis. Likely, this effect of alcohol can be explained by its ability to increase the permeability of cell membranes [46].

5.3.1.2.3 OTHER ASPECTS OF LIFESTYLE

As observations of epidemiologists show, some permanent household irritants can cause development of diseases that constitute a ground for the emergence of cancer. So, in those areas of Russia where very hot food (and, in particular, hot tea) is normal, esophageal diseases, which sometimes may turn into cancer, are observed more often than elsewhere.

The Soviet scientist S.P. Shilovtsev described the cancer of the tongue and oral mucosa in residents of some areas in Central Asia that occurs because of alterations in the tissues due to a tradition of applying "nas" under the tongue, which is a special blend of tobacco, ash, lime and oils. Similar observations were made in India, where there is an old custom to put within one's cheeks some leaves of the betel plant lubricated with lime and sprinkled with tobacco. In Colombia, smoking laundresses, while doing their job, hold cigarettes between their teeth with the smoking end pointing into the mouth to make sure ash does not fall into the laundry. Cancer of the upper palate is common among them [48].

It is well known that some ways of cooking, such as grilling meat, may represent a carcinogenic hazard. There is no reliable epidemiological data for this, but there are strong suspicions that kebabs, barbecue, shawarma, and other foods cooked on open fire increase the risk of cancer in humans. The risk is represented by so-called heterocyclic amines, which cause cancer in experimental animals.

These substances are generated from the amino acid creatinine under the influence of high temperature when frying on open fire or even in a very hot pan. They contribute to the development of tumors of the gastrointestinal tract, liver, and pancreas. However, in fried foods not only heterocyclic amines are dangerous, but fat dripping from meat onto the coals can also turn into toxic substances. Rising with the smoke, they can fall down back to the pieces of meat. In meat and fish cooked on open fire, the appearance of other carcinogens, polycyclic aromatic hydrocarbons (PAHs), is also possible.

In the summer of 1956, a group of research workers from the Division for the Study of carcinogenic agents of the Institute of Experimental and Clinical Oncology of the USSR (now Cancer Research Centre RAMS) undertook a special expedition to the Baltic Republics on order to study the role of smoked foods in Latvia. It turned out that the coastal population, mainly engaged in fishing and consuming relatively large amount of home-smoked fish, developed cancer of the digestive tract more often than the population of the inland areas involved in farming. Later (in 1959), the occurrence of malignant tumors of the digestive tract in workers of meat-smoking plants was studied. The workers undoubtedly ate more smoked food than the surrounding population. The cases of cancer of the digestive tract in this category of workers was also significantly higher.

5.3.2 Long-Term Experiments on Animals

As already mentioned, the epidemiological method for determining the toxicity and carcinogenic risks of chemicals is an involuntary experiment on humans. So, only the properties of some substances accidentally caught in the view of researchers can be determined. For systematic testing of a huge amount of substances suspected of genotoxicity, the experiments on animals have to be carried out.

The main advantage of this method is the appearance of a tumor, that is, the same phenomenon as with people. Biotransformation of carcinogens occurs in the body of a mammal that modulates with maximum precision the biochemical events that occur in the human body.

In testing for carcinogenicity, a lot of different animals are used. In some cases, justified by the danger of a tested substance, experiments were performed on gross animals such as horses, dogs, or monkeys. But such experiments are extremely expensive and require a long time (i.e., years) to obtain useful results. Therefore, the basic information available to date on the carcinogenicity of chemicals has been and still is to be gained with experiments on small animals, usually rodents. Mice and rats are the basic experimental animals.

During testing, it is necessary to obtain data for the description of the dependence "dose–effect" relationship and also to obtain the parameters of this relationship, allowing to establish the normative dose and concentration values. It is required to identify the primary target organ affected by a carcinogen. It is advisable to be able to extrapolate the obtained results for the possible effect of small doses of the agent on humans. Finally, it is desirable to obtain data in order to test the hypotheses of possible mechanisms of action.

A strict protocol is developed for carrying out tests. Usually, the test protocol is a standard approved by appropriate authorized organizations, which enables comparison of results obtained in different laboratories. Protocols vary depending on the type of substance, its carcinogenic activity, and the method of exposure. Here are offered some provisions of a typical protocol in accordance with the international standard, which give an idea of the complexity of the work.

First of all, it's essential to choose a method of body exposure to a substance that depends, of course, on how people get in contact with the substance at work or in the environment. One of three basic ways may be used: the intake of a substance with food or drink, inhaling the substance (i.e., the inhalation method), or administration of it through the skin (the experts call this method "dermal").

By the tests for toxicity, the LD_{50}, that is, 50% mortality value, is determined (see 4.4.4) which means a number of species members die during the experiment. By the tests for carcinogenicity, the value of a 50% probability of tumor induction is determined. In this case, in principle, all animals can survive by the end of the testing period and then they are slaughtered in order to get the data.

Typical test duration is 18 months for mice and hamsters and 24 months for rats. For certain types of animals with a longer life span and/or a lower susceptibility to tumors, the test may last 24 months for mice and hamsters and 30 months for rats. The test substance is administered to animals seven days a week—without any break for weekends or holidays.

The temperature in the experimental room for animals is going to be 22 °C (±3 °C), the humidity 50%–60%. The lighting is artificial, in a sequence of 12 hours' light and 12 hours' dark. Depending on the already-known or previously obtained information on toxicity and the supposed bioactivity of the substance, several doses are set. For each dose, a group of rodents is selected and there is always a control group. In testing on rodents, for each dose and in the control group, at least 100 animals are used (50 females and 50 males). In order to identify genotoxicity and to obtain statistically reliable data on low-active substances, many more animals may be required. Dosage levels and intervals between them must be selected so that it could be possible to detect the dose–effect dependence.

It is not possible here to describe in detail all the features of the test protocol. But in order to bring into focus all the complexity of this work, it has to be pointed out that a daily inspection of animals, control of consumption of food and drink, as well as weighing at least once a week should be carried out in the first 13 weeks. On completion of the test, the animals are slaughtered and a full histopathological examination of organs and tissues is conducted, with all tumors undergoing a more detailed microscopic study.

It is obvious that experiments on animals, even on small rodents, require the work of a skilled personnel and a lot of time, and the costs estimated at the end of the twentieth century were up to 2 million dollars per substance. Certain difficulties in carrying out experiments on animals are presented by activists of various animal protection societies. Besides, the data obtained in experiments on animals can but with certain reservations be used to estimate the intensity of the impact of the same carcinogens on humans.

The above-described difficulties of the experiments on mammals stimulated the use of unconventional animal species with high sensitivity to the carcinogenic effects and a shorter period of tumor development. In particular, an example of this approach can be performed by the use of certain types of aquarium fish, which were used for testing a large number of well-known chemical carcinogens for rodents and humans. The frequency of induced tumors in some kinds of fish is close to that of mammals, but their latent period of tumor development is much shorter. This makes this method considerably more economical. The methods for testing human carcinogens on "fish models" are widely tested and described in detail.

However, the use of the so-called short-term tests (STTs) for a faster identification of carcinogenicity can significantly simplify the solution of the problem of determining the carcinogenic risks of the chemicals.

Some results of determining carcinogenicity on animals are given in Section 5.4.

5.3.3 Physicochemical Methods

If molecules of certain substances can cause damage to DNA (and thereby facilitate the initiation of cancer) but other molecules do not do that, a certain difference could be expected between the molecular properties of both these molecules. If we could identify these properties, detection of carcinogens would become much easier. It would be reduced to the measuring of some physicochemical parameters that can be done quickly, cheaply, and even automated.

An overview of the early history of the physicochemical approach toward determining the biological activity of molecules was made by Lucas Christophorou in [49,50].

First of all, the physicists paid attention to a quite easily observed and more or less studied phenomenon—luminescence. It was found that a large number of luminescent organic molecules play an important role in biology. For example, carcinogenic activity of some members of a series of aromatic hydrocarbons has attracted great interest in the physical properties of this class of molecules. Bright luminescence is often characteristic for strong aromatic carcinogens (3,4-benzpyrene, 1,2-benzanthracene).

Attempts have been made to relate the energy of the optical transitions corresponding to fluorescence and/or phosphorescence, the position of the first excited state, the spectral shift and quantum yield of fluorescence, with certain biological functions of organic molecules. These attempts had, at best, limited success. Jones in his works in 1940, 1941, and 1943 investigated the ultraviolet absorption spectra of 370 aromatic hydrocarbons, both substituted and nonsubstituted. He aimed to tie bathochromic shift (a shift of the absorption spectrum into the long wave range) with substitution by an alkyl group, and the dependence of this shift on the position of the substitute with a carcinogenic index of methyl-substituted 1,2-benzanthracene. The greater shift was associated with more active molecules, but more definite conclusions could not be made. L. Christophorou makes an overview of a large number of attempts in order to find a correlation between the different luminescent characteristics of molecules and carcinogenic activity. He rightly points out that to know some individual physical properties of molecules is not enough. The fundamental biological problems of their toxicity, mutagenicity, and carcinogenicity cannot be solved this way.

The Nobel laureate Albert Szent-Gyorgyi paid attention to the important role of electrons in biological processes in the cell [51]. The cell has a rich source of electrons in the atoms N, S, and O transmitted during interactions. There is evidence that electrons in the cell exist in a quasi-free state and that they can be transported through the chain of carbon atoms with conjugated multiple bonds of the type -C=C-C=C-. On this point, see also [52]. The excited electrons can also move from one macromolecule to another over a distance of the order of 1.5–2 nm due to the tunnel effect. This all means that protein chains exhibit semiconducting properties. Many authors have drawn attention to the role played by the intermolecular electron transfer in the processes of biological regulation, protection, and cancer.

A most broad debate was sparked by the role of electrons and the charge transfer in the carcinogenic activity of organic molecules. A. Pullman and B. Pullman [53] studied the correlation between electronic donor–acceptor capabilities of aromatic molecules and their carcinogenic activity.

R. Mason [54,55] suggested that the initial stage of chemical carcinogenesis is the transfer of charge from the protein to the carcinogenic molecule. In these studies, the protein molecule is taken as the target molecule. The last step in this discussion was made by Elizabeth and James Miller [56,57], who showed that although chemical carcinogens differ significantly in their structure, they have one thing in common, and that is electrophilicity. This property is either characteristic of the direct carcinogen, or it is acquired by its metabolites.

It is fully determined (at least for the genotoxic molecules) that covalent binding of carcinogenic electrophilic metabolites with DNA sections that encode the enzymes controlling cell division, such as oncogenes or antioncogenes, is a key event in the initiation of chemical carcinogenesis. The transfer of an electron from the nucleophilic portion of the DNA molecule to the electrophilic molecule of a direct carcinogen is critical for such binding and for the formation of an adduct.

On this basis, it has been suggested that the identification of biologically hazardous molecules can be reduced to measuring of the molecule electrophilicity under study. Thus, it is necessary to measure a parameter that characterizes the electron capture by the test molecule. In all hitherto known attempts, the ratio of electron capture by the molecules dissolved in a buffered medium was measured. J. Lovelock with colleagues [58–60] used argon containing 1% hydrogen at a temperature up to 220 °C, as a buffer medium. The high temperature was required to make sure that the pressure of the tested substance vapor was sufficient to achieve the desired concentration.

In a series of works, George Bakale with colleagues [61] measured the rate of capture of electrons in a nonpolar liquid. The test substance was dissolved in liquid isooctane or cyclohexane, and the solution was injected into an ionization chamber. An impulse source of ionizing radiation created impulses of fast electrons with energy of 1 MeV and with duration of 15 ns that irradiated the solution in the chamber and created free electrons in it. Electrons moved in the chamber under the influence of an electric field and got captured by electrophilic molecules of the solved substance. Essentially free electrons, checking electrophilicity of molecules, performed the same role as the salmonella DNA molecules in Ames's test (see Section 5.3.4).

Based on the shape of the electric impulse in the chamber, one could determine the capture coefficient k_e. In Section 4.3, it was pointed out that for a molecule to capture an electron, they both should at least come into contact. Because their relative motion is diffusion, then, if nothing prevents the capture, the capture is called diffusion-controlled and the capture coefficient k_d is defined by the Debye–Smoluchowski formula (formulas 4.14

Table 5.2 Comparison of Predictive Criteria Based on the k_e Test and the Ames's Test [61]

Parameter	k_e Test (%)	Ames's Test (%)
Sensitivity	77	51
Specificity	90	82
Accuracy	85	69
Predictiveness	84	67

and 4.15 in Section 4.3). If the capture coefficient is less than this value, it means there is a thermodynamic barrier for the capture, and on this basis G. Bakale treated a substance with such capture coefficient as a noncancer substance. The boundary value of the capture coefficient was accepted by Bakale as being equal to the electron capture coefficient of molecules CCl_4: $k_d = 3.0 \cdot 10^{12}\ M^{-1}\ c^{-1} \approx 5 \cdot 10^{-9}\ cm^3/s$. So, if $k_e < k_d$, the substance is not a carcinogen, if $k_e \geq k_d$, the substance is a carcinogen.

Using this technique, Bakale tested a large number of substances where toxicity had already been studied in animal experiments, and mutagenicity by the Ames's test. In one study [61], G. Bakale offers results for 85 substances, including 35 carcinogens, 50 noncarcinogens, 27 mutagens, and 58 nonmutagens. Predictive criteria, calculated by comparing the data of the k_e test and the Ames test with indicators of carcinogenicity (see 3.2), are listed in Table 5.2.

It can be seen that the parameters of the k_e test, that is, determination of a single physical characteristic, turn out better than those of the famous and conventional biological test by Ames.

R. Benigni [62] had chosen a different way to implement the same idea. Instead of measuring the electron capture coefficient, he calculated the coefficient, sticking to the substance structural formula. Each structure group of the molecule was represented in the equation with a member with the corresponding coefficient, linear or quadratic, positive or negative. These coefficients were selected by comparing the calculated values with the experimental data on a rate of capture, measured by G. Bakale and coworkers [61].

5.3.4 Short-Term Tests, Bacterial Mutagenesis, Justification of the STT

5.3.4.1 Justification of the STT

Short-term testing (STT) is testing of chemicals for carcinogenicity that takes much less time than experiments on animal. STT is based on an assessment of the biological effects, correlated or causally related to carcinogenesis, and enables with a certain probability to predict the carcinogenic hazard of these or other factors. STT for identification of potential carcinogenicity are based on recent data on the mechanisms of chemical carcinogenesis. Quite a few different STTs have been assessed. This includes experiments on living organisms (in vivo), transformation of cells in tissue cultures (in vitro), induction of mutations in Drosophila flies, induction of tumors in aquatic organisms, and bacterial and cytogenetic tests. Among them, the most common and theoretically based are mutagenicity samples.

This is associated with the idea that the development of most tumors induced by carcinogens (at least at the initiation stage) is based on a genotoxic effect.

Substances entering the body through food, drinking, breathing or through the skin pass through numerous barriers and undergo metabolic transformations before hitting the target cell nucleus. Therefore, the tests in vitro in which the substance should not make this way and do not undergo these transformations might lead to results that are not quite reliable. In this sense, in vivo tests are much more reliable. Tests in vivo are much easier to extrapolate to humans.

Kirk Kitchin with colleagues from the Environmental Protection Agency conducted verification of a group of in vivo tests [63]. A group of halogenated hydrocarbons was selected (40 substances) with carcinogenic activity already proven in tests on rodents, and mutagenicity proven in the bacterial Ames's test (see below). According to a certain protocol, rats were fed by the test substance (the first dose was given 21 hours before slaughtering, the second, 4 hours before it). In 21 hours, the rats were slaughtered and their blood samples and liver tissues were tested. Biochemical analysis of samples enabled to identify the activity of certain enzymes that are produced in the body of the animal, which could signal the carcinogenic nature of test substance. The data obtained were subjected to statistical processing.

So, if the rats drank pure water, then after biochemical reactions with liver tissue, the hepatic ornithine decarboxylase activity manifested itself in the release of CO_2 in concentration 1.02 ± 0.2 nmol/g · hour. If 2-chloroethanol was added to the water, the release grew to 5.17 ± 2.1 nmol/g · hour. This result was acknowledged as statistically significant and the substance was declared a carcinogen.

The peculiarity of the experiments in vivo can be illustrated by the following example. The substance "monuron" (herbicide, chlorophenyl-dimethylurea) during the test on the content of hepatic cytochrome P450 was announced a carcinogen because the test result was 2.90 ± 0.28 nmol/g, which was significantly higher than the result in the control group 1.83 ± 0.28 nmol/g. At the same time, for the test substance 4-chloroacetylacetanilide, the result was 5.61 ± 0.30 nmol/g, which doesn't exceed the result in the control group (5.57 ± 0.63 nmol/g). Thus, this substance should be declared noncarcinogenic even though the value 5.57 is evidently higher than 2.90. Such rules are accepted in biology, because there exists no comprehensive understanding of the nature of biological processes in living organisms. Possibly, the initiation of the damage of genetic apparatus depends not only on the test substance but also on the season or some unaccounted factors, on the location of the planets, and so on. And so the testing result makes sense only in comparison with the result of the control group that was received simultaneously and strictly under the same conditions as the result in the experimental group.

When evaluating test results, one assumes that data obtained in experiments in vivo has more weight than the similar data obtained in vitro.

5.3.4.2 Ames's test

As a bacterial mutagenicity test, the most famous and widely used test of B.N. Ames has to be described.

Ames used a certain strain of Salmonella bacteria (*Salmonella typhimurium*) that as a result of mutation lost the ability to synthesize histidine for itself (biologists call this histidine auxotrophy). Therefore, in the control group, they do not grow on medium without histidine. The test substance is usually treated by a specially prepared fraction from rat liver for metabolic activation. To identify "direct" mutagens, the research can be also conducted without metabolic activation. Bacteria are treated with a test chemical and are incubated for a certain period. If the test chemical and/or its metabolites possess mutagenic activity, they will induce reverse mutations from auxotrophy to prototrophy for histidine, otherwise known as backward mutations. As a result, the bacteria recovers the ability to produce this amino acid and live on a nonhistidine medium. For the research, at least five different doses of the test compound have to be checked, and the necessary simultaneous control is required. In the control, the test compound gets subjected to treatment with an inert diluent such as distilled water.

The actual experiments with Ames's test, of course, are more complex than this brief description. Complications may be related, for example, to possible bactericidal power of the test substance, the need for repeated measurements, and so on. More information can be found, for example, in the works [64,65].

Most STT simulate individual stages of carcinogenesis. Therefore, the most successful is that using them as batteries, or sets, of tests. These batteries must meet several requirements. All STT included in a battery have to be complementary, that is, either vary in the final result (DNA damage, genetic mutations, chromosomal aberration, neoplastic transformation, disturbance of metabolic cooperation, etc.) or in the level of biological organization of the research subject (prokaryotes, eukaryotes, in vitro systems, in vivo systems). Thus, the sequence of tests involves the movement from simple to complex and from shorter to longer experiments.

In conclusion, we present the list of tests included in the minimum STT battery recommended by the Pharmacological Committee of the Russian Federation [66].

A. TESTS FOR DETECTING GENETIC MUTATIONS. Ames's Salmonella/microsomes test using exogenous activation of species with the S9 fraction of rat liver. The equivalent tests are: induction of recessive, sex-linked, lethal mutations or induction of somatic mutations in Drosophila.

B. CYTOGENETIC TESTS. Induction of chromosomal aberrations or micronuclei in bone marrow cells of mammals in vivo.

C. TESTS FOR DNA DAMAGE. Reparation test on *E. coli* or induction of bacterial cell SOS response. Induction of unscheduled DNA synthesis in mammalian cells or identifying DNA damage by the method of fluorometry or alkaline elution.

D. TESTS FOR PROMOTER ACTIVITY. Glomerular filtration rate test for violation of metabolic cooperation in a mixed culture of somatic mammalian cells.

E. DIRECT EXPRESS TESTS, IDENTIFYING CARCINOGENIC POTENTIAL OF TEST SUBSTANCES. Test for transformation of cells in culture or induction of tumors in aquatic organisms.

When evaluating test results, it is assumed that the positive effect has advantage over the negative. According to this rule, an integral positive result of the STT system is far more suggestive of potential carcinogenic risk of a specimen than a generally negative result is of its absence.

5.3.5 Correlation Between Structure and Biological Activity of a Molecule (QSAR)

The search for correlation between physicochemical properties and carcinogenic activity showed that a single physical property of a molecule cannot uniquely correspond to such a complex phenomenon as biological activity. However, this method may be useful in a battery of tests. Nowadays, a computer analysis has became available of large databases containing information about carcinogenicity, mutagenicity, toxicity, and other biological properties of thousands of chemicals from different classes. Besides, the physical and chemical characteristics of the molecules of many substances are sufficiently well known. On this basis, a research direction, quantitative structure–activity relationship (abbreviated as QSAR), was developed and it continues to actively evolve [67].

The carcinogenicity forecasts with QSAR are based on analyzing the shape of a molecule and its individual structures. The most significant are the indices of the molecular bonds, quantum parameters (the energy of the highest occupied orbital and the lowest unoccupied orbital, charges on different atoms, the electron densities, etc.). Modulating factors also are taken into account, including octanol–water distribution in the system (lipophily) and presence of certain metabolic sites that define possible ways of converting a compound in the cell. To analyze the obtained data, QSAR equations and the instructive programs CASE (Computer Automated Structure Evaluation) or Multi-CASE, which are combined with programs such as META. The latter are needed to account for probable metabolic transformations of the test compound. For each chemical series, there should be a sufficiently large and uniform training sample comprising compounds of this type, characterized by independent biological methods. This gives an opportunity to build forecasts, based on elements of similarity with known genotoxic carcinogens of a given chemical series, on a larger number of structure alerts than in the 60s. It has to be noted that even today, a satisfactory forecast of a compound's carcinogenicity based on QSAR is only possible for compounds with mutagenic action mechanism.

QSAR methods are commonly used in pharmacology with synthesizing of new drugs [68], in order to determine a relation of physicochemical characteristics of the molecules with not only their carcinogenicity but also their toxicity [69]. In a simple QSAR model, some toxicity characteristics (carcinogenicity, mutagenicity, etc.) are calculated. For example, it can be LD_{50}, by using a simple linear function of the physical characteristics of molecules, which are called molecular descriptors: x_1, x_2, x_3, ...

$$\text{Toxicity} = a_1 x_1 + a_2 x_2 + a_3 x_3 + \ldots,$$

where a_1, a_2, a_3 ... are coefficients selected by fitting to the known toxicity data. A detailed summary of descriptors used is provided on the site of the U.S. Environmental Protection

Agency (EPA) [69]. Recent information on the effectiveness of QSAR methods can be found in [70,71].

The overview of carcinogen screening methods in Section 5.3 shows the complexity of the problem. First, identifying carcinogenic hazards of substances is time- and labor-consuming, expensive, difficult, and, unfortunately, unreliable. Second, the detecting of substances in the environment with an already identified carcinogenic hazard requires the use of complex physical and chemical methods of analysis. There is no universal method suitable for the detection of any substance.

All this makes the fight against the danger of chemical carcinogenesis much more complicated than with radiation hazards.

5.4 Chemical Carcinogenesis Databases

5.4.1 Databases

Currently, much information has been accumulated on the toxicity, carcinogenicity, mutagenicity, and other manifestations of biological activity of different molecules. The information is summarized in databases available on the web for all interested. The databases are constructed differently. Some require requests for individual substances. In others, a review of the entire database is possible. In most cases, there is a detailed list of articles used as sources for the database. Within the databases, the required material may be searched for by at least its CAS number and name. Some databases may provide more opportunities. Here we give a brief overview of some databases. A more detailed review is in article [71]

CPDB: The Carcinogenic Potency Database [72]. This contains data on mutagenicity (Ames's test result) and carcinogenicity (experiments on mice and rats) for 1547 substances. The total number of examined substances is changing, naturally. Several years ago the author worked rather long with this database. Then it contained 1352 substances.

Danish QSAR database [73]: The Danish Agency for Environmental Protection (The Danish EPA) has developed a database containing information on 166,000 substances. This information does not contain any experimental information, but only predictive QSAR model-based data, obtained from a large number of different in vitro and in vivo STT, as well as in chronic experiments on rodents.

DSSTOX: Distributed Structure–Searchable Toxicity [74]. Developed by the National Center for Computer Toxicology (NCCT US EPA).

ECHA CHEM: European Chemicals Agency [75]. The database is a source of information on the chemicals manufactured and imported in Europe. It covers their hazardous properties, classification, and information on how to use them safely. This information can help to replace the most hazardous chemicals by safer alternatives. Database contains information about 12,735 unique substances (September 2014).

ESIS: The European Chemical Substances Information System (ESIS) [76]. Collection of data from several sources: European Inventory of Existing Commercial chemical Substances (EINECS); European List of Notified Chemical Substances (ELINCS);

No-Longer Polymers (NLP); the Biocidal Products Directive (BPD), PBT (Persistent, Bioaccumulative, and Toxic) or vPvB (very Persistent and very Bioaccumulative) assessments of Existing Substances; Classification and Labelling (C & L), the Export and Import of Dangerous Chemicals, High Production Volume Chemicals (HPVCs) and Low Production Volume Chemicals (LPVCs), including EU Producers/Importers lists; IUCLID Chemical Data Sheets; EU Priority Lists and EU Risk Assessments produced under the Existing Substances Regulation (ESR).

EXCHEM: [77] Developed in Japan by authorized organizations. Contains data on 250 substances. Much of the information is in Japanese, but some data is in English.

GAP: The Genetic Activity Profile Database [78]. Contains data on 300 substances, and about 200 STT.

IARC: The International Agency for Research on Cancer. Contains a large variety of information, links to it can be found at [79].

ISSCAN: Database of the Italian Instituto Superiore di Sanito in Rome [80]. Contains data on more than 1150 substances tested in chronic experiments on rodents.

NTP: The US National Toxicology Program [81]. Contains data on more than 500 two-year experiments on two types of rodents, about 300 toxicity studies of shorter duration and more than 2000 studies by STT method, in vitro and in vivo. In addition, you can find data on immunotoxicity, toxicity as regards to development and reproduction.

ToxRefDB [82]: Developed jointly by NCCT and the Department of Programs for Pesticide Studies. The database primarily contains information about the toxicity of pesticides, of about 330 substances.

TOXNET: Database of US National Library of Medicine (NLM) [83]. This is an effective cluster of different databases containing information on toxicology, hazardous substances, and options of toxicity manifestation. Among others, the CCRIS (Chemical Carcinogenesis Research Information System) and the GENE-TOX database with information on mutagenicity and carcinogenicity are included. CCRIS contains about 9000 results. The results of substance tests are commented by experts in the respective fields. GENE-TOX was developed by US EPA and contains information on genotoxicity of more than 3000 substances.

5.4.2 CPDB Description [72]

Here we offer a brief description of one of the databases. This permits not only to get acquainted with database material but also to highlight the problems of carcinogenicity studies.

At the time when the author worked with CPDB, it contained values TD_{50} for 1352 substances studied in chronic experiments in rats and mice. The database also contained the results of mutagenicity research by Ames's test in a two-digit form (+, –). Also, the preferred organ in which the tumor arose was specified. However, in this brief description, this issue is not included. The TD_{50} value range for rats and mice is shown in Table 5.3, and the total amount of mutagens, nonmutagens, carcinogens and noncarcinogens in Table 5.4.

Table 5.3 Values TD_{50} for Rats and Mice vary in the Following Range

	For Rats Substance	TD_{50} (mg/kg/day)	For Mice Substance	TD_{50} (mg/kg/day)
Minimum	2,3,7,8-Tetrachlorodibenzo-*p*-dioxin	$4.57 \cdot 10^{-5}$	2,3,7,8-Tetrachlorodibenzo-*p*-dioxin	$1.56 \cdot 10^{-4}$
Maximum	SX purple	$2.45 \cdot 10^{4}$	C.I. pigment red 3	$3.55 \cdot 10^{4}$

Note: The smaller the TD_{50}, the more toxic is the chemical.

Table 5.4 Quantities of Mutagens, Nonmutagens, Carcinogens, and Noncarcinogen in CPDB

	Ames's Test	Rats	Mice
Mutagens for Salmonella, carcinogens for rats and mice	360	516	391
Nonmutagens for Salmonella, noncarcinogens for rats and mice	393	537	479
Not investigated	598	292	477

The maximum values of TD_{50} are very close to the limit value of the amount of matter that the animal can potentially consume. Actually, an adult rat consumes 30–50 g of feed per day. If to assume the weight of the rats as ~ 200 g, then TD_{50} for the substance "SX purple" is approximately of 5 g/day. But apart from the poison, a rat should also receive fats, proteins, carbohydrates and vitamins. If to assume that the poison should not exceed 10% of the daily diet, it turns out that TD_{50} is just what can be actually put into food. This means that there may be carcinogenic substances, the doses of which are even bigger. However, they just cannot be verified in experiments on animals.

Analysis of the data tabulated in CPDB shows that there are substances proven as carcinogens in mice tests, but safe for rats and vice versa. Furthermore, there are substances which in Ames's Salmonella test performed mutagenicity, but showed no carcinogenicity in tests on one or even both groups of rodents, and vice versa. Using the methods of tests verification described in Section 3.2, one can calculate the predictive criteria. The results are shown in Tables 5.5 and 5.6. Tables are constructed such that all calculation components are obvious. For example (Table 5.6), 276 mice carcinogens were tested on rats. Only 202 of them showed coinciding properties, that is, were tested positively (TP). Hence the sensitivity of the test on mice for determination of carcinogenicity in rats is 0.73. The remaining numbers the reader can analyze himself.

Table 5.5 The Ability of the Ames's Test to Predict Carcinogenicity for Rats and Mice

Predictive Criteria	Chemical Carcinogenicity for Rats	Chemical Carcinogenicity for Mice
Sens = TP/N_C	222/339 = 0.65 ± 0.03	172/290 = 0.59 ± 0.03
Spec = TN/N_{nC}	215/314 = 0.68 ± 0.03	208/296 = 0.70 ± 0.03
Acc = (TP + TN)/N	(222 + 215)/(339 + 314) = 0.67 ± 0.03	(172 + 208)/(290 + 296) = 0.65 ± 0.03
P.pr = TP/(TP + FP)	222/(222 + 99) = 0.69 ± 0.1	172/(172 + 88) = 0.66 ± 0.1
N.pr = TN /(TN + FN)	215/(215 + 117) = 0.65 ± 0.1	208/(208 + 118) = 0.64 ± 0.1
P-3	0.30 ± 0.03	0.27 ± 0.04

Table 5.6 The Ability of Experiments on Mice to Predict carcinogenicity for Rats and of Experiments on Rats to Predict Carcinogenicity for Mice

Predictive Criteria	Test on Mice Predicts Carcinogenicity for Rats	Test on Rats Predicts Carcinogenicity for Mice
Sens = TP/N_C	202/276 = 0.73	202/280 = 0.72
Spec = TN/N_{nC}	238/316 = 0.75	238/312 = 0.76
Acc = (TP + TN)/N	440/592 = 0.74	440/592 = 0.74
P.pr = TP/(TP + FP)	202/280 = 0.72	202/276 = 0.73
N.pr = TN /(TN + FN)	238/312 = 0.76	238/316 = 0.75
P-3	0.405	0.405

After reviewing these extracts from the database on carcinogenic and mutagenic substances, one can make following conclusions.

First, the range of TD_{50} is very wide: 8–9 orders of magnitude.

Second, lots of substances turn out to be carcinogens and mutagens. In a rather large group of substances in this database, selected on the basis of industrial and research needs, more than a half turned out to be carcinogenic for rats (516 of the 1060 tested) and slightly less than a half turned out carcinogenic for mice (391 of the 875 tested). There is still no definite answer to the question why so many substances are carcinogenic. The author of the famous bacterial test, B. Ames, believes that it is the effect of the large doses that are used in the tests (for more detail, see Section 6.6) [84].

Third, there are quite a lot of substances that are carcinogenic for mice and noncarcinogenic for rats and vice versa. But people probably differ from rats and mice greater than these two kinds of rodents differ from each other. This means that the test on rodents can be used to predict human carcinogenicity only with great reservations.

So, the problem still exists. And the solving of this problem has been keeping biologists busy for a long time.

References

[1] Chemical Abstracts Service (CAS). www.cas.org.

[2] Walters D, Grodzki K. Beyond limits? Dealing with chemical risks at work in Europe. Elsevier; 2006. p. 416.

[3] National Library of Medicine. National Institutes of Health. Toxnet: Toxicology Data Network. http://www.nlm.nih.gov/pubs/factsheets/toxnetfs.html.

[4] Environment–Risk–Health. Toxicological and identification data bases. http://erh.ru/dbchemicals.php. [in Russian, contains a lot of references to the U.S. and International databases].

[5] Russian registry of potentially hazard chemical and biological substances. http://www.rpohv.ru. [in Russian].

[6] National Toxicology Program. http://ntp.niehs.nih.gov/results/dbsearch/index.html.

[7] Toxicology Data Network. http://toxnet.nlm.nih.gov/cgi-bin/sis/htmlgen?HSDB.

[8] ITER: International Toxicity Estimates for Risk. http://toxnet.nlm.nih.gov/cgi-bin/sis/htmlgen?iter.

[9] Carcinogenic Potency Database. http://toxnet.nlm.nih.gov/cgi-bin/sis/htmlgen?CPDB.htm.

[10] Bennet PB. Inert gas narcosis. In: Bennet PB, Elliot DH, editors. The physiology and medicine of diving. London: Saunders; 1993.

[11] Smith BW, Monthioux M, Luzzi DE. How does xenon produce anesthesia? Nature 1998;396:324.

[12] Statistics – National Statistics. Drug misuse: findings from the 2013/14 crime survey for England and Wales. Updated 15 August 2014. https://www.gov.uk/government/publications/drug-misuse-findings-from-the-2013-to-2014-csew/drug-misuse-findings-from-the-201314-crime-survey-for-england-and-wales.

[13] Web-Atlas: environment and the health of Russian population. http://www.sci.aha.ru/ATL/ra00.htm. [in Russian].

[14] State report about the conditions of environment in Russian Federation in 1995, Moscow, CMP, 1996, 458 p. [in Russian].

[15] Study: air pollution causes 200,000 early deaths in US. http://www.voanews.com/content/air-pollution-linked-to-early-death/1739804.html.

[16] Green Living. Air pollution statistics. http://greenliving.lovetoknow.com/Air_Pollution_Statistics.

[17] U.S. Environmental Protection Agency. http://www.epa.gov/airquality/urbanair/

[18] US Environmental Protection Agency. Air pollution emissions overview. http://www.epa.gov/airquality/emissns.html.

[19] Pauling L, Robinson AB, Teranishi R, Cary P. Quantitative analysis of urine vapor and breath by gas-liquid partition chromatography. Proc. Natl. Acad. Sci. U. S. A. 1971;68:2374–6.

[20] Dweik RA, Amann A. Exhaled breath analysis: the new frontier in medical testing. J. Breath Res. 2008;2. (3): 0300301

[21] Russian Register of Potentially Hazardous Chemical and Biological Substances. Rospotrebnadzora of Russia. http://www.rpohv.ru/security/. [in Russian].

[22] MedlinePlus. A service of the U.S. National Library of Medicine. National Institutes of Health. http://www.nlm.nih.gov/medlineplus/ency/article/002435.htm.

[23] Consumer Rights Protection Society. http://ozpp.ru/consumer/useful/article5.html. [in Russian].

[24] U.S. FDA. Frequently asked questions. http://www.fda.gov/Food/GuidanceRegulation/FSMA/ucm247559.htm

[25] Calton M, Calton J. Rich food poor food: the ultimate grocery purchasing system (GPS). Primal Blueprint Publishing; 2013.

[26] Goyanes C. 14 Banned foods still allowed in the U.S. On the site SHAPE. http://www.shape.com/blogs/shape-your-life/13-banned-foods-still-allowed-us.

[27] U.S. Food and Drug Administration. Food additives & ingredients. http://www.fda.gov/food/ingredientspackaginglabeling/foodadditivesingredients/ucm2006845.htm.

[28] Food and Drug Administration. www.fda.gov.

[29] The alliance for a healthy tomorrow. www.healthytomorrow.org.

[30] Dodson RE, Nishioka M, Standley LJ, Perovich JGB, Rudel RA. Endocrine disruptors and asthma-associated chemicals in consumer products. Environ. Health Perspect. 2012;120(7):935–43.

[31] Westervelt A. Study highlights hidden dangers in everyday products – even the "green" ones. http://www.forbes.com/sites/amywestervelt/2012/03/08/study-highlights-hidden-dangers-in-everyday-products/.

[32] Watts DL. Trace elements and other essential nutrients. Clinical application of tissue mineral analysis. Meltdown Intl., 1995

[33] Järup L. Hazard of heavy metal contamination. Br. Med. Bull. 2003;68:167–82.

[34] Supplements-And-Health.com. Evaluating the health risks from dietary supplements. http://www.supplements-and-health.com/health-risks-dietary-supplements-2.html.

[35] Hodgson E. A textbook of modern toxicology. John Wiley and Sons; 2010.

[36] Gadaskina ID, Tolokontsev ID. Poisons – yesterday and today: essays on the history of poisons. Leningrad: Nauka; 1988. http://n-t.ru/ri/gd/yd.htm [in Russian].

[37] Leenson I. Arsenic and human's health. http://www.krugosvet.ru/node/42947?page=0,0. [in Russian]

[38] Lin X. Contents of arsenic, mercury and other trace elements in Napoleon's hair determined by INAA using the k0 method. Analytical application of nuclear technique. Vienna: IAEA; 2004. p. 159–162. http://www-pub.iaea.org/MTCD/Publications/PDF/Pub1181_web.pdf.

[39] Bolozdynya AI, Obodovskiy IM. Detectors of ionizing particles and radiations. Principles and applications. Dolgoprudny: Publ. House Intellect; 2012. p. 204 [in Russian].

[40] U.S. EPA. Evaluating and eliminating lead-based paint hazards. http://www2.epa.gov/lead/evaluating-and-eliminating-lead-based-paint-hazards

[41] The alliance for a healthy tomorrow. http://www.healthytomorrow.org/toxics/chemicals.html

[42] Committee of Nutrition. Aluminum toxicity in infants and children. Pediatrics 1996; 97: 413-6. http://pediatrics.aappublications.org/content/97/3/413.

[43] Taylor G. It's not just the mercury: aluminum hydroxide in vaccines. Adventures in Autism (March 9, 2008). www.adventuresinautism.blogspot.com.

[44] Smart RC. Chemical carcinogenesis. In: Hodgson E, editor. A textbook of modern toxicology. 3rd ed. Wiley; 2004.

[45] Chaklin AV, Journey in the wake of mystery, Mysl, 1967; Chaklin AV, Journey in the wake of mystery continues, Mysl', 1980. [in Russian].

[46] Zaridze DG. Epidemiology, mechanisms of carcinogenesis and cancer prevention. The talk on the III congress of oncologists and radiologists of UIS, Minsk, 2004. http://health-ua.com/articles/989.html. [in Russian].

[47] Zaridze DG. Every young fool ever tried to smoke. http://www.akzia.ru/tema/02-10-2009/2646.html.

[48] Ageyenko AI. Cancer, Transcript of malignancy (early diagnosis, treatment, rehabilitation). Niola Press, 2007, 128 p. http://scrubsquared.net/18304-rak-rasshifrovka-zlokachestvennosti-a-i-ageenko.html. [in Russian].

[49] Christophorou L. Biophysics and bioelectronics (Chapter 9). Atomic and molecular radiation physics. J. Wiley & Sons; 1971. p. 601–12.

[50] Christophorou L, Hunter SR. From basic research to application (Chapter 5). In: Christophorou L, editor. Electron-molecule interactions and their applications. Academic Press; 1984. p. 401–12.

[51] Szent-Györgyi A. Bioelectronics: a study in cellular regulations, defense and cancer. Academic Press; 1968.

[52] Rubin AB, Shinkarev VP. Electron transport in biological systems. Nauka; 1984. [in Russian].

[53] Pullman B, Pullman A. Quantum biochemistry. Interscience Publishers; 1963. xvi + 867 p..

[54] Mason R. Electron mobility in biological systems and its relation to carcinogenesis. Nature 1958;181:820–2.

[55] Mason R. Role of electron and exciton transfer in carcinogenesis. Radiat. Res. Suppl. 1960;2:452–61.

[56] Miller JA, Miller EC. Chemical carcinogenesis: mechanisms and approaches to its control. J. Natl. Cancer Inst. 1971;47:V.

[57] Miller EC. Some current perspectives on chemical carcinogenesis in humans and experimental animals. Cancer Res 1978;38:1479.

[58] Lovelock JE. Affinity of organic compounds for free electrons with thermal energy: its possible significance in biology. Nature 1961;189(4766):729–32.

[59] Lovelock JE, Zlatkis A, Becker RS. Affinity to polycyclic aromatic hydrocarbons for electrons with thermal energies: its possible significance in carcinogenesis. Nature 1962;193(4815):540–1.

[60] Lovelock JE, Simmonds PG, Vandenheuvel WJA. Affinity of steroids for electrons with thermal energies. Nature 1963;197(4864):249–51.

[61] Bakale G, McCreary RD. A physico-chemical screening test for chemical carcinogens: the $k_e$test. Carcinogenesis 1987;8(2):253–64.

[62] Benigni R. QSAR prediction of rodent carcinogenicity for a set of chemicals currently bioassayed by the US National Toxicology Program. Mutagenesis 1991;6(5):423–5.

[63] Kitchin K, Brown J, Kulkarni A. Predicting rodent carcinogenicity of halogenated hydrocarbons by in vivo biochemical parameters. Teratogen Carcin. Mut 1993;13:167–84.

[64] Maron DM, Ames BN. Revised methods for the Salmonella mutagenicity test. Mutat. Res 1983;113:173–215.

[65] Ames BN, McCann J, Yamasaki E. Methods of detecting carcinogens and mutagens with the Salmonella/mammalian-microsome mutagenicity assay. Mutat. Res. 1975;31:347–64.

[66] Belitsky, GA et al. The forecast of carcinogenicity of pharmacological agents and auxiliaries in short-term tests. http://www.medline.ru/public/fund/pharmcom/7.phtml. [in Russian].

[67] Devillers J. Evaluation of the OECD (Q)SAR Application Toolbox and Toxtree for predicting and profiling the carcinogenic potential of chemicals. SAR QSAR Environ Res 2010;21(7–8):731–52.

[68] Richon AB, Young SS. An Introduction to QSAR Methodology. http://www.netsci.org/Science/Compchem/feature19.html.

[69] Cronin MTD, Dearden JC. Review QSAR in toxicology. 1 Prediction of aquatic toxicity. Quant. Struct. Act. Relat. 1995;14:1–7.

[70] Fjodorova N, Vracko M, Novic M, Roncaglioni A, Benfenati E. New public QSAR model for carcinogenicity. Chem. Cent. J. 2010;4(Suppl. 1):S3. http://www.ncbi.nlm.nih.gov/pmc/articles/PMC2913330/.

[71] Serafimova R, Gatnik MF, Worth A. Review of QSAR models and software tools for predicting genotoxicity and carcinogenicity. Joint Research Center, European Commission, Institute for Health and Consumer Protection; 2010. 52 p. http://ihcp.jrc.ec.europa.eu/our_databases/jrc-qsar-inventory/review-qsar-models.

[72] CPDB The Carcinogenic Potency Database. http://potency.berkeley.edu/cpdb.html.

[73] Danish QSAR database. http://ecbqsar.jrc.it/.

[74] DSSTOX Distributed Structure-searchable Toxicity. http://www.epa.gov/ncct/dsstox.

[75] ECHA CHEM: (ECHA – European Chemicals Agency). http://echa.europa.eu/chem_data_en.asp.

[76] ESIS: The European chemical Substances Information System (ESIS). http://ecb.jrc.ec.europa.eu/esis/.

[77] EXCHEM. http://dra4.nihs.go.jp/mhlw_data/jsp/SearchPageENG.jsp.

[78] GAP: The genetic activity profile database. http://www.ils-inc.com.

[79] IARC: The International Agency for Research on cancer. http://www.iarc.fr/.

[80] ISSCAN: Data base of Instituto Superiore di Sanito в Риме. http://www.iss.it/ampp/dati/cont.php?id=233&lang=1&tipo=7.

[81] NTP: The US National Toxicology Program. http://ntp.niehs.nih.gov.

[82] ToxRefDB. http://www.epa.gov/ncct/toxrefdb/.

[83] TOXNET: Database of US National Library of Medicine – NLM. http://toxnet.nlm.nih.gov.

[84] Ames BN, Gold LS. Paracelsus to parascience: the environmental cancer distraction. Mutat. Res. 2000;447:3–13.

6 Radiation and Chemical Hormesis

6.1 Definition of "**Hormesis**": Arndt–**Schulz** Law

To a first approximation, our world is linear. In the basic laws of nature—Newton's law, Ohm's law, Hooke's law, and many others—there is a linear relationship between impact and result. Acceleration is proportional to force; electric current is proportional to the voltage; distortion is proportional to mechanical stress, and so on. Deviations from linearity only occur with increasing intensity of exposure. Thus, disharmony in oscillatory processes, shock waves and vortices in flows of gas, water, electrons, and so on, arise. However, in the case of living matter, the situation with linearity is completely different. Obvious nonlinear effects appear even at low intensities when various factors affect the living substances. In the second half of the nineteenth century, the German scholars E. F. W. Pflüger, R. Arndt, and H. Schulz formulated what was later declared as the general biological law, Arndt–Schulz law, according to which the weak stimulations evoke activity in living elements, the medium ones increase it, the strong ones inhibit it, and the very strong ones paralyze it. In other words, depending on the intensity of exposure, the effect changes its sign. To be more precise, we point out that this law is also manifested in some effects in the nonliving complex physicochemical systems.

This section discusses the application of this law to the impact of both chemicals and ionizing radiation on living organisms. A specific example of the manifestation of this law is homeopathy, discussed further in Section 6.4.

Stimulating, beneficial effects on body systems by an external agent, which with large doses exhibits a damaging (toxic) effect, is called hormesis. According to some authors, the term was introduced by C. Southam and J. Ehrlich in 1943. According to other sources, it was proposed as early as in the 1920s [1].

It should be noted that various researchers gave different names to the complex dose–response dependence typical of hormesis. In scientific literature, one can come across the terms *paradoxical effect, inversion phenomenon, bimodal curve,* and others. But by now *hormesis* has become the accepted term.

The idea of chemical hormesis appeared in the late nineteenth century, when dozens of experiments had been described with the positive impact of various metals and chemical compounds on the growth and development of plants, fungi, algae, and bacteria [2].

Thus, it was known that at high doses these substances inhibit the development of organisms.

The term "radiation hormesis" was proposed in 1980 by T. D. Luckey and it implies a beneficial effect of ultra-low radiation doses.

To date, the ideas of hormesis are taken quite seriously by the scientific community and they are developing intensively. In his article of 2010, one of the major authorities in this field, E. Calabrese, pointed out that up to then more than 1200 articles on hormesis had been already published. More than 80% of these articles were published after 2000 [3]. The concept of hormesis is included in the programs of the major textbooks on toxicology and appears to be the main content of special issues of magazines and the main theme of scientific conferences. The May 1987 issue of the magazine *Health Physics* [4] is entirely devoted to the problem of hormesis. Hormesis has become the subject of several scientific studies [5–8].

The hormesis problem is closely related to the problem of threshold or linear dose–response dependence in stochastic manifestations of exposure to damaging factors. Questions like which effects are related to the phenomenon of hormesis and which aren't, and whether it is time regulatory agencies started to consider the phenomenon of hormesis in guideline documents, became the subject of intense debates [9]. It can be said with confidence that the concept of both chemical and radiation hormesis is becoming one of the basic biological concepts.

Manifestations of hormesis have been observed in medicine, molecular biology, pharmacology, nutrition, agriculture, microbiology, immunology, toxicology, aging, physiology, and carcinogenesis [10], that is, in almost all branches of biology.

Some researchers, while not denying the stimulating effect of small doses, are actively loath to consider this effect as favorable [11,12]. Increasing fertility or productivity of agricultural crops and animals can be useful to humans, but harmful to the actual crop or animal. In some cases, the concepts "harmful" and "useful" can be interchanged. For example, a drug for cancer chemotherapy may be effective at higher doses when it produces an inhibitory effect on cell proliferation, but harmful at low doses when it stimulates proliferation and consequently tumor growth [13].

So, the most important achievement in the study of low-dose effects is the revelation of nonlinearity of the dose–response dependence and the understanding that exposure to substances and radiation in low doses cannot be measured by a simple extrapolation of the experimental data obtained while using high damaging doses [14].

If negative impact of the damaging factor is plotted on the ordinate, then one might say that the hormesis curve has the J-shape, as shown in Figure 6.1A. If the positive effect is plotted, the graph has the shape of an inverted "U," as shown in Figure 6.1B.

6.2 The Definition of "Low Doses"

The impact of small doses of both radiation and chemicals is a separate area of research that is very important, because most of the population is exposed to small doses. It is extremely interesting, because it can provide answers to fundamental scientific questions about the mechanism of action of radiation and chemicals on living organisms. The topic of low doses is currently discussed in detail in scientific circles and political arenas, among civilians and ecologists. This topic is a field of desperate struggle among different points of view. Based on these discussions, crucial decisions on protection of the population and the environment are taken.

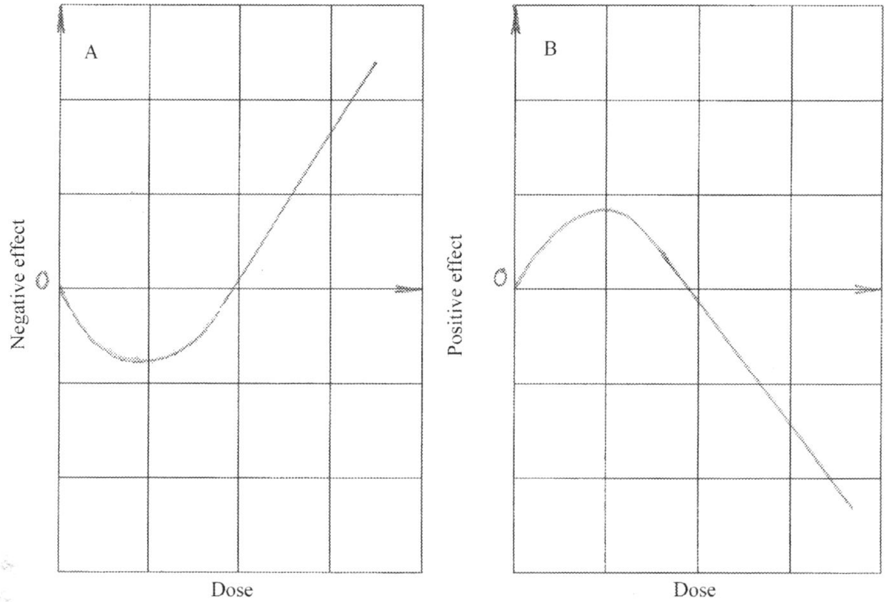

FIGURE 6.1 The J-shape variant (A) and the inverted U-shape variant (B) of hormesis dose–response dependency

Currently, a number of authors are inclined to think that at low doses the mechanisms of radiation and chemical exposure are rather close and that the response of biological systems to weak radiation and chemical exposure demonstrates their similar properties (see, e.g., [15]). Therefore, it seems appropriate to combine all the information on the low-dose effects in one chapter.

For the sake of accuracy, it should be noted that a significant percentage of researchers do not believe that low doses generally deserve to be highlighted; that there are no special effects of low and especially ultra-low doses and that everything is linear from zero up to the very high quantities.

Information on "low doses" can be found in a number of official documents of international and national organizations [16,17], in some articles and books [18–23], and in some magazines [24], and the latest information on the problem of low doses is available in the web material by A. N. Koterov [19] and S. Gilbert [25].

Before describing low-dose effects and, in particular, the effects of low-dose radiation on human health, it is useful to define what is a low dose and what is the upper limit of the range that can be referred to as a low dose. Naturally, the question of the limits for radiation and chemical exposure (because of the difference in the measuring units) is fundamentally different, despite the proximity of their mechanisms. Let us first look at radiation effects.

It is obvious in advance that the boundary of small doses may not be precisely defined. It is different for different subjects, for the young and the elderly people, for the healthy and the weak ones. It is going to depend on the radiation effect under consideration and on the damage that was taken into account. For example, a dose of 10 Gy is absolutely lethal to a human but low and not causing any damage to some radioresistant reptiles, while it is conducive to the growth and development of mustard seeds. For the same organism a dose is going to be small or large depending on biological criteria. A dose of 1 Gy for humans will be low if the death of the organism is taken as a damage criterion but, at the same time, high enough to cause a sharp increase of chromosomal aberrations in blood lymphocytes [26,27]. Nevertheless, the analysis of the limit value for low doses and low dose rates is carried out and such limits are getting set. At least two approaches are possible here—physical and biological.

According to physical notions, the upper limit for the range of "low doses" and "low-dose rates" can be estimated as based on the considerations of microdosimetry. If the damage to the cell nucleus depends on energy release in the nucleus and the interaction between the cells can be ignored, then the deviation from linearity is observed if at least 2 tracks pass through the main sensitive target—the nucleus. Taking into consideration the statistical nature of the interaction of particles with the substance, one can expect that the low probability of two particles passing through the core corresponds to the absorption dose of 0.2 mGy (from low linear energy transfer [LET] radiation to gamma radiation ^{60}Co, for the typical tissue and a spherical cell nucleus with a diameter 8 μm). Thus, one particle will pass through only 18% of cells, and more than one particle will pass through less than 2% of cells.

Microdosimetry criterion for determining the limits of the low doses range is applicable to biological effects, where the energy released in the cell causes damage only to this cell. It can be applied to the process of cell death, mutations, and chromosomal aberrations.

Application of this criterion to cell transformation and development of cancer is less obvious. It requires a modification when, for example, the probability of the effect is affected by several particles hitting the cell sequentially, like in the multistage carcinogenesis model, or when the development of an appropriate radiation effect requires interaction between the cells, for example, as assumed in the bystander effect.

The criterion discussed for a low dose depends essentially on the dimensions of the sensitive target. The above-selected diameter of 8 μm corresponds to the size of the nucleus of some typical cells but, in principle, it can be bigger or smaller. Furthermore, if only a part of the nucleus autonomously responds to the action of radiation, the sensitive volume may be lower and, accordingly, the limit of the energy value higher. For the diameter of 8 μm, the obtained limit of a dose value is 0.2 mGy, with a nucleus diameter of 4 μm the dose limit is 0.8 mGy and with a diameter of 32 μm the dose limit will be about 0.01 mGy. On average, one particle passes through the nucleus with the absorbed dose of 1 mGy.

The limit of a dosage rate is determined taking into account the possible DNA repair and its dependence on time. It turns out that the limit value of the dose rate is about 10^{-3} mGy/minute. If we take as a basis the requirement that only one particle should pass through the cell during the lifetime of an organism (assumed 60 years), which virtually excludes tracks intersection, we obtain the limit value of the dose rate of the order of 10^{-8} mGy/minute.

When irradiation is conducted with high LET particles, fluctuations in the release of energy increase dramatically. Usually 0.1 mGy gamma radiation is equated to 300 mGy of radiation with alpha particles. At a dose of 1 mGy from alpha particles, the particles will pass through only 0.3% of the cell nuclei; the remaining 99.7% will not be affected by radiation. When a particle passes through the nucleus, it leads to the release in the nucleus of a very large dose, about 370 mGy on the average. At the same time, an individual nucleus can get any energy up to 1000 mGy.

A limit range of low doses can also be obtained on the basis of biological considerations, and such considerations may be several. First, one can start from the approximation of the observed dose–response dependence.

In the linear-quadratic model (see Section 4.4.4), the degree of cell damage is associated with a dose according to the ratio

$$I(D)=\alpha D+\beta D^2, \tag{6.1}$$

where α and β are constant coefficients, which vary for different types of cell lesions. This equation very well describes the induction of chromosomal aberrations in human lymphocytes and some other effects. For certain types of chromosomal aberrations in lymphocytes, the ratio α/β is about 200 mGy at gamma radiation by a ^{60}Co source. The ratio α/β corresponds to an average dose at which both linear and quadratic terms make equal contributions to the biological effect. We can assume that the relationship will be purely linear at a dose 10 times lower, that is, 20 mGy. At this dose, the quadratic term is only 9% of the total effect. Even at 40 mGy, the quadratic term gives only a 17% input. On this basis, the range limit of low doses can be considered as approximately 20–40 mGy.

Other approaches are based on a study of the effect of radiation on animals or epidemiological studies on contingents of people who survived the bombardments of Hiroshima and Nagasaki. For calculation details, the reader is referred to [28] or to the reports of the United Nations Scientific Committee on the Effects of Atomic Radiation (UNSCEAR [16,29,30]).

Despite the considerable volume of information indicating the presence of a threshold in the action of ionizing radiation on human health, the authorized organizations cannot as yet take the final decision for the establishment of new rules and regulations. Therefore, in declarative documents, the linear nonthreshold dose–effect dependence is used, but for the range of low doses, the coefficient that reduces the role of radiation is introduced. In various studies, this coefficient is called a dose and dose rate effectiveness factor (DDREF), a dose-rate effectiveness factor (DREF), a linear extrapolation overestimation factor (LEOF), or a low-dose extrapolation factor (LDEF) [16].

Based on the above considerations about small dose criterion, UNSCEAR in its report in 1993 [30] concluded that for the evaluation of the risk of cancer in humans at irradiation with low LET radiation, the DDREF should be used at doses of less than 200 mGy. For high LET radiation, such an amendment should not be applied.

It should be noted that some researchers have introduced the concept of "ultra-low doses." One of the variants of such an approach is as follows: the low doses represent the zone of obvious stochastic effects and the ultra-low doses are those at which no effects (even stochastic) can be reliably observed. Russian radiobiologist Prof. E.B. Burlakova, and colleagues, conducts long-term research on the reactions of living systems to ultra-low impacts. The group suggests to define the ultra-low doses of radiation as doses, starting with which the effect changes its sign, that is, with the transition from inhibition of cell growth to its stimulation [18]. As to the effects of chemicals, in the same paper [18], it is proposed to consider ultra-low doses of biologically active substances as the doses, the effectiveness of which cannot be explained from the currently accepted positions and requires the development of new concepts.

Burlakova et al. [18], analyzing some of the published works and the studies of their own laboratory, specify 10^{-11} M = $6{\cdot}10^{9}$ cm^{-3} to 10^{-13} M = $6{\cdot}10^{7}$ cm^{-3} as a limit value of "ultra-low doses" on the basis of data on the number of cellular receptors and affinity of ligands to them, or estimating the number of biologically active molecules per cell.

6.3 Radiobiology Paradigm

After more than a hundred years of accumulating vast factual material on the damaging effects of ionizing radiation on living organisms, the basic mechanisms have been securely identified and theories of radiation damage have been created. As a result, a frame of views was established called the paradigm of radiobiology. A brief summary of this paradigm could be formulated as follows [31]:

1. radiation is harmful;
2. radiation is harmful at any dose;
3. there are no low-dose effects that are not predictable on the basis of the known effects of exposure to high doses.

According to this paradigm, which for a long time has been supported by the majority of radiobiologists, there is no "low-dose" problem and to identify the limit of this range is meaningless. The only thing that seems sensible is to identify the minimum dose at which it would still be possible to detect damage by using certain techniques. It was assumed that even natural radioactive background is surely harmful. It just fundamentally cannot be eliminated. Therefore, it determines the lower limit of the damaging effects of radiation on all living organisms.

However, data have been accumulated proving that the effects of low doses actually exist gradually. In addition, the general biological considerations, based on the similarity of radiation and chemical exposure mechanisms and already well-proven manifestations of the Arndt–Schulz law (see Section 6.1) of exposure to chemicals, required a revision of the basic assumptions of the paradigm.

In the second half of the twentieth century in three independent laboratories in France, in the USSR and in the USA, in experiments on various biological objects, it has been determined that natural background radiation is essential for the normal growth and for the development of a living organism. In particular, in the laboratory research of A. M. Kuzin [26] on higher and lower plants and growing infant rats, it was experimentally found out that if to create a shielded volume with a reduced, compared to natural radiation, background, one can observe the slowing down of metabolic processes, cell division, growth and development of the organisms. The same results were obtained in 1967 in the H. Planel laboratory [32,33] in France on Drosophila eggs and in the Argonne Laboratory in the USA by T. Luckey [8,9,34–36]. If to restore the exposure by introducing a radiation source into the volume, the slowing of the processes in the same experimental conditions would completely stop [27]. Reduction of the weekly ionizing component of the natural radiation background of the Earth by a factor of 20 resulted in an increased rate of aging and dying of yeast cell strains.

This is an important experimental confirmation of the adaptation of all the living to small doses of radiation, at least at the level of natural background radiation.

6.4 Chemical Hormesis

The phenomenon of chemical hormesis is well known and its existence is quite certain, though the term "hormesis" itself is less known.

In the Section 5.1 it was already mentioned that there are no (or almost no) impeccably toxic substances. Most of the substances that are called poisonous can in certain doses be harmless, useful or even necessary for normal development and for functioning of the body. Toxicity of arsenic and antimony in large doses and their medicinal properties in small doses is well-known. Another example is trace elements (see Section 5.1.8). In low concentrations, these elements are vital to the body, but in high concentrations they can cause poisoning.

An interesting example is oxygen toxicity. In frogmen and scuba-diver practice, it is well known that oxygen at partial pressure above the normal (0.21 ATA) can have toxic effects on the body. Oxygen poisoning is caused both by the magnitude of the partial pressure and the exposure time. Excess oxygen causes an increase in the amount of oxidized hemoglobin and decrease of reduced hemoglobin. It is the deoxidized hemoglobin that carries out the carbon dioxide transportation and the reduction of its content in the blood leads to a delay of carbon dioxide in the tissues and finally to hypercapnia. Hypercapnia manifests itself in short breath, facial flushing, headache, seizures, and finally, loss of consciousness.

With excess of oxygen, its metabolism in tissues changes. During reduction of oxygen molecules, at least partially, intermediate products and free radical oxygen forms appear. Free radical metabolites possess high activity and act as oxidants that damage biological membranes. Lipids—the main component of biological membranes—are extremely easily oxidized compounds. Free radical oxidation of lipids often becomes a branched chain reaction, prone to self-maintenance, even after normalization of the oxygen amount in the body. Many products of this reaction are themselves highly toxic compounds and can damage biological membranes.

Informations on chemical hormesis were published with regard to bacteria, phytoplankton, algae, pollen germination, and more complex plants. Substances that were found out to display hormesis effects include heavy metals, chlorinated hydrocarbons, insecticides, substances inhibiting the growth of plants, the hydrocarbons of the crude oil, and antibiotics [37].

Many issues of the *Biological Effects of Low Level Exposures (BELLE)—Newsletter* are devoted to chemical hormesis [24]. For example, v. 16, № 1, 2010, was devoted to the subject of "Hormesis and homeopathy"; v. 14, № 3, 2008, was titled "Hormesis and Ethics"; and v. 12, № 1, 2004, was "Economics and Hormesis," and so on.

Here are some examples. In order to inhibit plant growth, that is, to make plants shorter, more compact, and more attractive for commercial use, special substances have been developed, for example, phosphon. However, its use for regulation of the growth of peppermint (*Mentha piperita*) led not only to growth inhibition while using normal doses ($>\sim 10^{-7}$ M) but also stimulated growth at lower doses.

Another example is related to chloroform. It is known to be carcinogenic for mice and rats and also as a potential carcinogen for the humans. However, it was found that low doses of chloroform (~60 mg/kg/day) increased the life expectancy of laboratory animals by approximately 10%, relative to the control group.

Well known in global practice is the usefulness of moderate consumption of alcohol, caffeine, or nicotine. At the same time, large doses of these substances can be very dangerous or even deadly.

During the Second World War, penicillin, which had only recently been discovered, saved a lot of lives. But penicillin then was scarce and doctors used small doses, trying to conserve it. It appeared that at low doses, penicillin had the opposite effect; not only did it not suppress but it stimulated the growth of Staphylococcus.

The greatest specialist in pharmacology and toxicology, Ed. Calabrese and colleagues, having evaluated more than 50,000 dose–effect dependencies, concluded that hormesis dependencies are much more common than threshold dependencies, let alone the linear dependencies. Linear or threshold models describe the results of measurements less precisely than a hormesis model [38,39].

Dose–response dependency with hormesis dominates in all areas of biomedical sciences, including toxicology and pharmacology. Dependence with hormesis becomes a universal model, regardless of the biological model, of the measured effect and of the chemical and physical properties of substances, the impact of which is analyzed.

The most extensive chronic experiment on animals to date, which included more than 24,000 rodents, clearly demonstrated the presence of hormesis in a dose–response dependency in the research of bladder cancer.

Nevertheless, regulators, defining the standards for the use of various substances, apply the linear model or the threshold model, leaving out the latest achievements.

6.4.1 Homeopathy

In the sixteenth century, the German alchemist and physician Theophrastus Bombastus von Hohenheim (1493–1541), known as Paracelsus, defined the rule of action of toxic substances on humans as follows: "Alle Dinge sind Gift und nichts ist ohne Gift; allein die

Dosis macht, dass ein Ding kein Gift ist." In English it reads: "Everything is poison, and nothing is devoid of toxicity; only a dose makes the poison unnoticed." This formula is often presented in a simplified form of "dose makes the poison" or "All is poison, all is medicine; both are defined by the dose."

Ideas of Paracelsus are embodied in the works of the homeopathy founder Samuel Hahnemann (Christian Friedrich Samuel Hahnemann, 1755–1843). The doctrine of the homeopathic effect of medicines was first described in 1796, and in 1810 Hahnemann's essay "Organon of Rational Healing Art" ("Organon der Rationellen Heilkunst") was published, being the first systematic description of the new doctrine. Research and application of the new method continued, as reflected in the six-volume edition of "Absolute Pharmacology" ("Reine Arzneimittellehre"), published in 1811–1819.

Having carried out experiments with many medicinal substances, Hahnemann came to the conclusion that any one of them causes in a healthy human the symptoms of the disease, which it was supposed to treat. From this, the principle "similia similibus curantur," that is, "like cures like," was derived. Small doses of drugs have a curative effect if large doses cause symptoms similar to those of the disease. This principle seems to be quite reasonable and scientifically sound, especially in the light of the Arndt–Schulz law formulated later (see Section 6.1) and the current data on chemical hormesis.

However, since the therapeutic effect was caused by low doses, another principle appeared on the basis of homeopathy: "the stronger the dilution, the more effective the medicine." Low doses of homeopathic medicines were prepared by dilution of normally prepared solutions. The prepared solution was diluted several times in a 10-fold (d = decimal) or a 100-fold (c = centesimal) proportion. The result was attomole (10^{-18} mol/L), zeptomole (10^{-21} mol/L), and even lower concentrations. In one of the guides on homeopathy, the author came across an indication of a 30c dilution. This means 30 consecutive dilutions by 100 times each. The trivial calculations of the number of molecules of the active ingredient in the final portion of a homeopathic medicine show that in the thus-prepared medication, it is unlikely to find at least one molecule of the original substance. This assessment has now become commonplace in any criticism of homeopathy. Of course, not all homeopathic medicines are based on such strong dilutions, but also weaker dilutions leave very little of the active ingredient in the resulting medicine.

Classic therapy methods were still purely empirical, and in most cases not very effective, at the time when the widespread occurrence of homeopathy received, and surely deserved, recognition. Homeopathy was based on a principle that already existed since the time of Hippocrates and was supported by a certain theoretical basis.

To justify the homeopathy of the past, it should be said that in the times of its formation not much of molecules and of their quantities was known. Only in 1811, Amadeo Avogadro suggested that a mole of any substance contains the same number of molecules regardless of its nature. But the actual number was determined only in 1865 by Loschmidt. As you know, Avogadro's number is equal to $6.022 \cdot 10^{23}$. This number determines the so-called Avogadro limit ($\sim 10^{-24}$ mol/L). Ultra-molecule or ultra-low dilutions are sometimes called BRAN (beyond the reciprocal of Avogadro's number).

Why now, at the present state of knowledge, the ideologists of homeopathy still insist on strong dilutions is hard to comprehend. Perhaps, for the same reason why the orthodox Jews still do not eat pork? Once, the Lord, feeling pity for His Peculiar People, knowing that the pork in a hot climate deteriorates quickly, which leads to serious poisoning, imposed the taboo on it to be eaten. But now He, all-seeing and all-knowing, can't help but witness the possibilities of refrigerating, so why can't he give some hint to lift the ban. But so far He does not. Or maybe He does, but people just don't understand.... Anyway, the formerly rational ban has acquired the halo of holiness and worship. May the homeopaths also take with some religious reverence those greater dilutions.

It has to be said that modern homeopaths spend considerable efforts for the scientific rationale of the therapeutic effect of strong dilution. Various mechanisms were imagined, according to which water structure may be changed under the influence of the molecules present in it before. Water is considered to remember the images of these molecules, and these images are considered to produce a healing effect. Alternatively, hypothetical, nonexistent clusters, nano-bubbles, nanoparticles, and other similar formations are made up.

If to discard this fantastic line of homeopathy, in all other respects it is a quite reasonable and apparently quite useful branch of medicine. However, it still has to get rid of the mark of unscientific character.

One can't but take into consideration the proximity of the concepts of hormesis and homeopathy. Of course, they are not the same thing. Homeopathy is a therapeutic method, and hormesis is a biological phenomenon. Homeopathic treatment is aimed at accelerating and facilitating the normal healing process. Hormesis is characterized by overcompensation, which homeopathy has never claimed. However, there is much of a common ground between them, the evidence of which is, for example, a special issue of the *BELLE Newsletter*, called "Hormesis and homeopathy" [40].

6.5 Radiation Hormesis

As the low-dose range is the field of a particular interest and because the studies of the effect of low doses on humans can only be performed in random populations exposed to the low doses, though higher than the normal natural radiation background, a lot of work is being done to clarify the low-dose effects on bacteria, plants, fungi, tissue cultures, and various animals (mammals). Since the formation of the concept of "low dose," so much information has been accumulated that to present any detailed analysis of these materials within one book is really not possible. The interested reader is referred to the papers of official organizations, to the monographs and articles in scientific periodicals [16,17]. Here, we are going just to try and find in the published materials the answers to two questions. Is a threshold of stochastic radiation effect observed in the dose–effect dependency (and how reliable is the observation)? Is there a reason to believe that, at doses below the exposure threshold, the radiation effect changes sign, that is, produces beneficial effects on the body? In other words, does radiation hormesis exist?

Let us specify the range of doses that have to be considered in relation to these issues. If we exclude from natural background about 1.2 mSv, which account for exposure to light alpha particles due to inhalation of radon and its decay products, then the average natural radiation background with low LET radiation is about 1 mGy/year. We assume that by now it has been reliably proven that radiation of such intensity is either harmless or even beneficial.

The upper limit of the dose rate threshold for deterministic effects is tens of Gy/year, let's assume ~10 Gy/year. Thus, a possible threshold of stochastic effects for the humans may have a range of 1–10^4 mGy/year.

6.5.1 Experiments on Bacteria, Plants, Fungi, and Tissue Cultures [17]

In the period of early development of radiobiology as the main type of radiation effect, the induction of chromosome aberrations by radiation was mainly studied, while those damages could be easily and securely recorded by the instruments used in radiobiological experiments at that time. In the 1970s, a technique was developed for quantitative study of specific gene mutations in cultures of somatic cells. Postradiation genomic instability, bystander effect, and other nontarget effects were subsequently found that offered researchers new tools. But it also set new questions on the impact of these phenomena on radiation carcinogenesis.

The main attention was paid to the analysis of the dose–response dependence and the relation between the efficiency of damage induction and LET [17]. The researchers have tried to advance into the low-dose range as much as they were allowed by experimental equipment and statistics. In most work, the known linear-quadratic dependence was observed (see Section 4.4.4). At low doses, of course, the linear component dominated. In a study of induction of chromosomal aberrations, the linear dependence was observed at low doses down to ~20 mGy of low LET radiation. Below this value, statistical provision was insufficient to exclude the theoretical possibility of dose threshold.

Because of special characteristics of the experimental equipment, analysis of gene aberrations produces much more accurate results and allows one to move to the doses that are lower than the registration of gene mutations [17]. Therefore, typical values of the lower limit of the dose dependences are ~100–200 mGy. Just in one case with particularly sensitive systems, it was managed to get to ~10 mGy. Nevertheless, there were studies in which one was able to detect a low-dose threshold. So experimental data were approximated in [17, p. 86, paragraph 57] by the function of the type

$$I = C + a(D - D_{th}),$$

where D_{th} is the threshold. Assuming the values $C = 0.0013$ and $a = 0.040$, the best estimate of the threshold offers: $D_{th} = 9.7 \pm 4.5$ mGy (±1 standard deviation, $\chi^2 = 4.0$ for 5 degrees of freedom). If this result is taken as valid, it is clear that a reliable threshold registration is only possible if the dose–response measurement is done with cell cultures at doses considerably lower than 10 mGy.

An extensive review of the published results on the effects of radiation hormesis can be found in the book of A.M. Kuzin [27]. Here, some information from his book are offered.

There is a lot of evidence obtained from experiments on animals and plants, that small doses stimulate cell proliferation. Life activity and fertility of animals increase, their health status improves, life expectancy lengthens, and irradiation of pre-sown seeds increases yield.

Effects of radiation hormesis were observed under the action of ionizing radiation on animals and vegetating plants. At doses up to 50 cGy in mammals, chickens, and fish, the stimulation of fertility was observed and an increase in survival and in growth rate. In plants, acceleration of growth processes, more intense branching, and stimulation of the development of the generative organs occurred.

Back in 1976, G. T. Chaffey et al. [41] reported that chronic exposure to total accumulated doses of 10, 20, and 80 cGy enhances the immune response of the T-lymphocyte system in the spleen of mice. Proliferation of T-lymphocytes increased by 15%, 40%, and 60%, respectively, in relation to a nonirradiated control group.

Stimulation of cell division as the effect of low doses of ionizing radiation is observed in a number of other biological subjects: mammalian cell culture, blue–green algae, and ciliates. A single irradiation dose in this case was in the range 0.01–0.5 Gy and dose rates during chronic exposure did not exceed 5 cGy/day.

Despite the uncertainty of the situation with the threshold, there is no doubt about the lower efficiency of damaging radiation effect at low doses and at low dose rates. It is believed that this is due to increased efficiency of reparation of premutagenic damage at low dose rates. This is usually taken into account by introducing special lowering coefficients called DDREF or DEF (see Section 6.2).

There are serious reasons to believe that the induction of double breaks of DNA chains is the main mechanism for creating chromosomal aberrations. It is well known that one chain break can be easily repaired in a few hours. Rupture of two chains, in principle, can also be repaired, but the probability of a full repair of such an extensive damage is rather small. Even in the case where the cell retains viability, it will contain a mutation. Accumulation of such mutations in the cell leads to the development of malignant tumors. Usually, double breaks are done by particles with high LET radiation. However, double breaks can also be produced by low LET radiation by successive passages through the cell of several (two) particles [42]. It is experimentally shown that the ratio of the number of single breaks and DNA base damages to the number of double breaks is 50:1. The chance of double breaks per cell is about four per 100 mGy [43]. Based on these concepts of the mechanism of cell damages, a threshold in the dose–effect dependence is only possible in case of a single passage of the particles through the cell.

6.5.2 Experiments on Animals

Special characteristics of experiments on animals, the experimental advantages, and the methodological disadvantages of such experiments are discussed in Section 5.3.2. Here one only has to mention that the lower the dose and the lower the sensitivity of an animal, the more species members are required to obtain a statistically secure result.

UNSCEAR experts cite the following calculations. If a mice breed is taken with a very low frequency of spontaneous leukemia ($1 \cdot 10^{-4}$) and high sensitivity to the induction of leukemia by radiation (about $1 \cdot 10^{-1}$ Gy^{-1}), it is sufficient to use about 300 irradiated animals and the same number in the control group to record a significant increase ($p = 0.05$) in induction of tumors at a dose of ~100 mGy for the whole body. For recording the effect from the dose of 10 mGy, groups of 4000 animals are required.

If one uses a mice breed with a higher frequency of spontaneous disease ($7 \cdot 10^{-3}$) and a lower sensitivity to radiation ($7 \cdot 10^{-3}$ Gy^{-1}), in order to fix a noticeable increase in the likelihood of induction of acute leukemia at a dose of 100 mGy, now $1.2 \cdot 10^5$ animals are required, which is unrealistic.

With low-dose experiments on animals, in a large number of studies the evidence of threshold existence has been found. A short overview of these results on the basis of official publications is offered here [16,17]. Interested readers will find in these reports references to the original publications.

The extensive studies on mice (about 18,000 animals were used with acute exposure to gamma-quantum of ^{137}Cs at 0.45 Gy/minute) enabled to detect the dose–effect relationship (leukemia) of the type (6.1), with $D_{th} = 0.22 \pm 0.14$ Gy, $\chi^2 = 2.6$ with 5 degrees of freedom.

The process of thymus tumor induction has undergone quite intensive research. In most cases, a threshold could be observed. Thus, J. R. Maisin and colleagues found out that a dose–effect curve begins to be proportional to dose only at a dose of 4 Gy and more. This is clearly seen in Figure 6.2 [16]. R. L. Ullrich and J. B. Storer, exploring the induction of lymphoma of the thymus in mice, approximated the experimental results with a linear-quadratic function, but its linear part at low doses had a very small, maybe even zero, slope, which also indicates the possible existence of a threshold.

Induction of ovarian tumors in mice irradiated with X-ray or gamma radiation also indicates the existence of a threshold. Approximation of curve data of type

$$I = 2.2 + 2.3(D + D_{th})^2$$

enables determination of the threshold value $D_{th} = 0.12$ Gy.

In [16 t is a table Table X on the page 105], a table is presented with values of the minimal doses at which one can observe an increase of cancers in mice in a large number of experiments. The number of minimal doses is about 0.5–1 Gy, and only in a few studies they are 0.16, 0.2, and 0.25 Gy.

It was found that in some cases, especially at internal irradiation by incorporated radionuclides, the time required for tumor formation at low doses significantly increases so much that it can exceed the lifetime of the animal. Perhaps, this very fact simulates the dose-dependence threshold. For example, in a group of mice irradiated at a dose of 0.7 Gy, the average life expectancy was 871 ± 105 days, which is significantly greater than in the control group (790 ± 144 days). At higher doses, the life expectancy decreased. It is noted

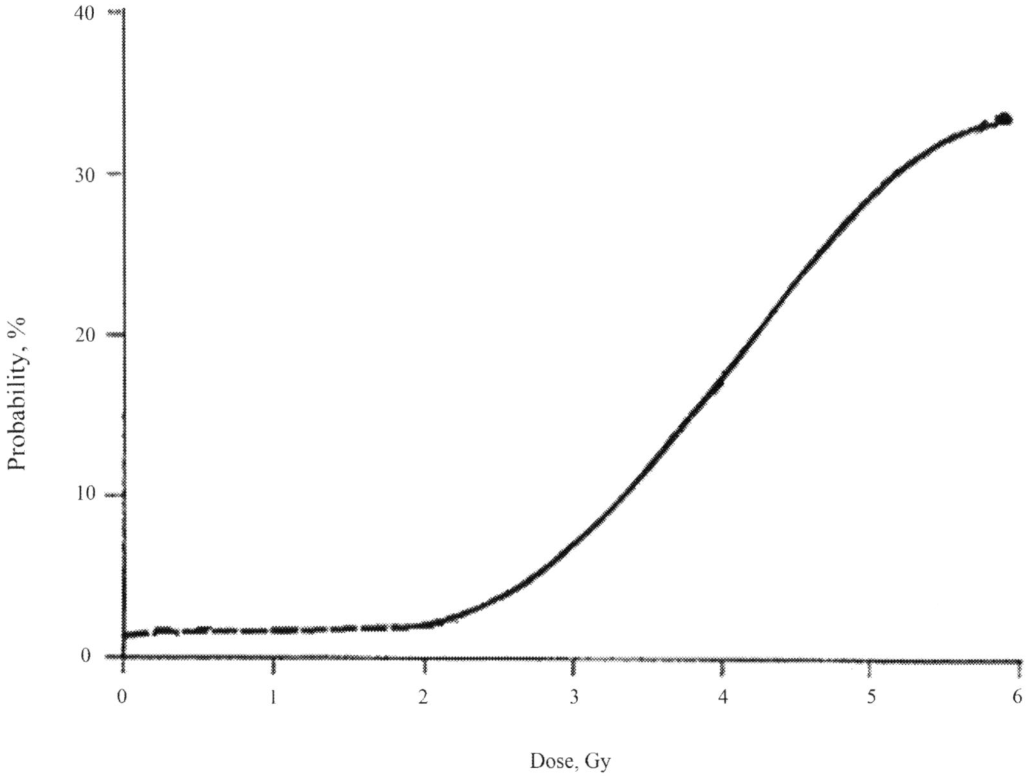

FIGURE 6.2 Induction of thymus lymphoma as a function of the dose at irradiation by X-ray quanta. Based on [17].

that at lower doses (<1 Gy), mainly benign tumors (adenomas) occur, at doses >1 Gy, they are mainly malignant.

Summing up the results of a large set of studies, the experts from UNSCEAR conclude that the lowest dose for which one can observe a definite increase in the number of tumors is 100–200 mGy for low-LET radiation. Large values of the minimum dose that result from some studies are attributed by experts to low sensitivity of test specimens, high induction frequency in the control group, or a small number of animals.

Noteworthy is the fact that threshold values in animal experiments are much higher than in the experiments on bacteria or tissue cultures. This fact, as well as some others, suggest that repair capabilities of a whole organism are higher than that of its parts.

6.5.3 Epidemiological Studies

It has been repeatedly stated that deliberate experiments on humans are not possible. However, since humanity has been faced with ionizing radiation, a large number of population groups accumulated who once received or have been receiving for a long time the radiation doses that can be defined as "low doses."

The mere enumeration of cohorts shows the scale of the data from which one can get the information about the effects of low doses on human health.

1. The survivors of the atomic bombing of Hiroshima and Nagasaki [44,45, p. 141–154]. Although the effects of radiation on humans has been studied for more than a hundred years and a large number of cohorts were formed that received excessive exposure for whatever reasons, yet the most detailed and important and giving, perhaps, the most extensive information, are the studies on the effects of nuclear explosions in Hiroshima and Nagasaki.
   Before atomic bombing, the estimated population of Hiroshima was 360,000 and of Nagasaki 250,000. A uranium atomic bomb (Little Boy) with energy ~16 kt was dropped on Hiroshima on August 6, 1945. Then on August 9 there was the second bombing with a plutonium bomb (Fat Man) with energy ~21 kt on the city of Nagasaki. Approximately 50% of the bomb energy was released in the form of blast, ~35% in heat and ~15 % in radiation. Immediately after the bombing and within the first 4 months, approximately 140,000 (38% of the whole population) in Hiroshima and ~70,000 in Nagasaki were killed or had died.
   Approximately 400,000 people in the two cities were still alive after the bombing. Soon after the war, a survey was undertaken of persons exposed in the course of the bombing and who survived. The survivors of the bombings are called *hibakusha*, a Japanese word that literally translates to "explosion-affected people." It was determined that among the survivors there are approximately 54,000 people that during the bombing were within <2.5 km from the hypocenters and received an increased dose and yet ~40,000 in the range of 2.5–10 km that received low doses. At present, the cohort of atomic bomb survivors who are enrolled in the program of Life Span Study (LSS) consists of 86,611 persons.
   Other study cohorts were added later, including individuals exposed in utero and children who were conceived after the bombing by irradiated parents—in total, approximately 200,000 individuals, 40% of whom are still alive today.
   People in Hiroshima and Nagasaki were exposed not only to radiation released directly from the bombs that lasted <1 min but also to radiation from radioactive fallout contained in black rain and from neutron activation in soils that can last much longer. Estimation of individual dose is a very complex problem. It depends on distance from the hypocenter, the shielding conditions, and the personal conditions: body size, position at the moment of explosion, following activity, and so on. Estimation of individual doses a couple of times were specified. The last estimation has a name DS02. The last update to the summary of the situation with the survivors is the report 14, 1950–2003 [46], wherein the authors fitted the cancer mortality data to an excess relative risk (ERR) model, concluding that there are no thresholds in dose-ERR dependence. This conclusion is subject to justifiable doubts as to Doss et al. [47], who believe that the conclusion is based on the wrong model selected. The authors evaluated the same data in another way and came to the conclusion that the accepted model does

not show a monotonic increase in ERR from zero dose, but becomes monotonically increasing only after the dose reaches approximately 0.27 Gy. The authors do not insist that such a threshold does exist, but they believe that research papers do not provide evidence to suggest that there is no threshold. However, in the other paper [48] analyzing the same data, the author changed the handling of the data and received good reason to believe that hormesis in the dependence of ERR on dose is observed. The result of this work is shown in Figure 6.3.

Thus, the cohort of atomic bomb survivors of LSS consists of 86,611 persons. Up to 2003, those who died of all causes reached 58%, that is, about 50,000 persons. Among those, approximately 12,000 died of cancer [46]. It is senseless here to point out exact figures. Time is ongoing and the figures change. The number of dead persons sharply rises with the increase in dose. There is no doubt that doses greater than 0.1 Sv are very dangerous, and analyses of the data confirm it. But with low doses, the situation

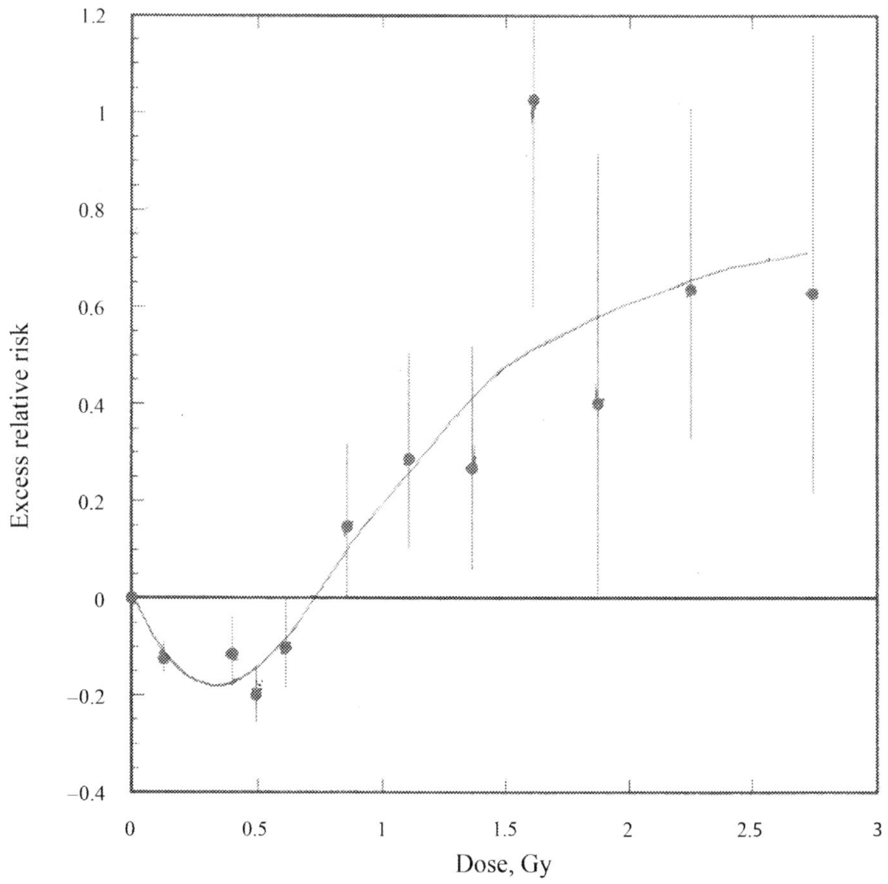

FIGURE 6.3 Excess relative risk (ERR) for all solid cancer in atomic bomb survivor. Based on [48], with the permission.

is not yet clear. It is quite possible that doses lower than 0.005 Sv are safe after all or even beneficial.

It is interesting that among survivors, there are double survivors, the persons who suffered the effects of both bombing. One of them is T. Yamaguchi [49]. He has been officially recognized by the government of Japan as surviving both explosions. He died of stomach cancer on January 2010 at the age of 93.

2. The employees of nuclear industry enterprises of the USSR and Russia, in particular, at the Mayak plant [44,45].

   Strictly speaking, the condition of the workers at the Mayak plant cannot be attributed to the section "low doses." External exposures and exposures of Mayak workers to plutonium far exceed those of other nuclear worker cohorts discussed in this chapter. For example, for the nearly 11,000 monitored workers hired before 1959, the mean cumulative external dose was 1.2 Gy, more than an order of magnitude higher than for the other cohorts [45, p. 201], and for the whole cohort the mean cumulative external dose was 0.8 Gy. Nevertheless, we present here some results about the Mayak cohort, because it offers a unique opportunity to obtain reasonably precise estimates of risks from medium- to high-dose *protracted* external exposure that can then be compared to estimates based on acute exposure, such as those obtained from A-bomb survivors.

   Approximately 21,500 people worked at the Mayak nuclear complex between 1948 and 1972. There were the employees of the nuclear reactor (4396), of the radiochemical plant (7892), the plant for the production of plutonium (6545), and two subsidiary factories (2724) of water treatment plants and a mechanical repair plant. The study of the irradiation conditions and cancer mortality shows that the workers of plutonium and radiochemical plants had a significant contribution of internal exposure by aerosols containing plutonium. For about a third part of the cohort, the contamination was measured. It was equal to 2.1 kBq.

   As a whole, for 40 years, 7067 persons died, 1730 from solid cancer and 77 from leukemia. For all solid cancers ERR = 0.15 Gy^{-1} (90% CI: 0.09, 0.20) [45, p. 201]. For leukemia, the estimated ERR/Gy was 6.9 (90% CI: 2.9, 15) 3–5 years after exposure and 0.5 (90% CI: 0.1, 1.1) 5 or more years after exposure. The estimate based on the entire period was 1.0 (90% CI: 0.5, 2.0).

   The studies show that a linear dose–response relationship fits the data best of all, as the doses are NOT low and a threshold or reduced effect at low doses is possible.

3. The population of the South Urals (region of the Techa River).

   The radiation catastrophe in Southern Urals is well known mainly because of the Kyshtym accident, named after the nearby town on the map. On September 29, 1957, there was an explosion of a vessel with a volume of 300 m^3 that contained the highly active waste of plutonium production. The whole activity of the waste was 740 PBq (where peta = 10^{15}) (20 MCi), approximately 10% of it was thrown into the air and then into the floodplain of the Techa River.

The tragedy, however, had begun much earlier. Starting from 1949, the radioactive waste had been disposed in the Techa River. It is believed that during the period 1949–1956 a total of 110 PBq (2.75 MCi) was dumped into the Techa River. About 95% of all active waste was allowed to flow into the river from March 1950 to November 1951.

In total, on the site contaminated as a result of the accident of 1957, there were 217 settlements with a population of about 270,000 people.

The population of the Southern Urals near the Techa River and Lake Karachay was exposed to both external and internal radiation, as they consumed fish, plant and animal products, and drinking water from the irradiated area. According to information received by UNSCEAR experts [50], the observations of the three groups of residents of 22 settlements with a population of 7852 who received an average of 49.6, 12.0, and 4.0 cGy showed within the next 30 years a significant ($p < 0.05$) decrease in mortality from various types of tumors in all three groups. Mortality was, respectively, 28%, 39%, and 27% compared to the nonirradiated population. The comparison of populations groups exposed by this accident to chronic inhalation of ^{239}Pu and having received 0.343, 1.18, and 4.2 kBq showed that the risk of lung cancer was significantly reduced compared to nonirradiated controls up to 0.56, 0.59, and 0.83, respectively.

4. Residents of Kazakhstan (and partially of Altai) who were exposed to radioactive fallouts from nuclear weapons tests at the test site in Semipalatinsk [44, paragraphs 109–110; 52].
5. The employees of enterprises of the nuclear industry of United States, United Kingdom, and Canada, a total of 95,673 employees, of whom 85.4% are male. The survey also included workers of 15 countries of 154 nuclear facilities, a total of 407,391 people, whose exposures to X-ray and gamma-radiation were proven by using personal dosimeters.

   In the paper by Gribben et al. [51], it was noted that the leukemia mortality rate was significantly ($p < 0.01$) reduced in a cohort of nuclear industry workers in England and the United States (130,000 people), compared to unexposed staff. Mortality from cancers and leukemia among workers of the nuclear industry in Canada was also 58% lower than the national mortality rate from these causes [52].

   The doses received by workers of the nuclear industry enterprises of 15 countries were determined on the basis of data from personal dosimeters. In most cases, mortality from all causes, in particular all types of cancer, was significantly lower than in the control group [17]. As one of the explanations, the authors of the study suggest that maybe the healthiest people were selected to work at the nuclear plants.
6. The workers and population exposed to uranium. In this group, uranium miners are not considered. Epidemiological studies of this population contain information on more than 120,000 workers.

   For workers involved in production of uranium, no link between cancer cases and type of work was found. In this group (120,000 people), 7442 cases of cancer were recorded, while the expected number for such a group was 8178 cases. No increased risk of cancer was found among people living near uranium processing plants or among individuals who used drinking water with increased content of uranium [44, paragraphs 126–130].

7. Population and staff irradiated during the testing of nuclear weapons in foreign countries.
   Approximately 500,000 people were exposed during the nuclear tests. The decrease in mortality from cancer was registered in the military observers at nuclear explosions in the atmosphere in the United States (32,000 people) and England (22,000) [17, p. 212–215; 53]. Mortality in Canadian military observers was 88% of the control group, while mortality from leukemia was 40% [54].
8. Populations inhabiting areas near nuclear facilities [17, p. 208–212].
9. U.S. Radiologists [44, p. 204–205].
   A cohort of 146,022 U.S. radiologists, of whom 106,884 were women, participated in radiologic activity during 1926–1982. Among the radiologists who worked prior to 1950, the risk of lung cancer and leukemia increased significantly, but in subsequent periods when exposure was reduced and more reliable methods of protection and control were introduced, the frequency of deaths from all types of cancer became less than could be expected for the population in general. None of the types of cancer showed any additional risk [17, p. 208–212].
10. Radiologists in China [44, paragraphs 117–120]
    The studied cohort consisted of 27,011 radiologists and technicians working in 1950–1980 in 24 provinces in China. The control group consisted of 25,782 people, who, working in health care institutions, didn't work with X-rays. In Chinese radiologists, the probability of all types of cancer was increased. There has also been a significant additional risk of leukemia. Like among the U.S. radiologists, the risk of cancer at early stages among people working with radiation was higher [44, paragraphs 117–120].
11. Long-distance flight crews and astronauts [17, p. 204, 44, paragraph 121].
    No apparent connection has been detected between mortality from all types of cancers and the flight duration of the crews engaged in long-distance flights at high altitudes. Moreover, there is some indication of a negative dependence of risk for all types of cancers with an increase in dose.
12. Patients undergoing radiation diagnosis and therapy [17, p. 155–188; 44, paragraphs 122–125].
    Strictly speaking, radiation therapy procedures are not included in the category "low doses." Usually, with radiation therapy very high doses are used, of dozens of Gy, but locally. Such a dose is lethal if applied to the whole body.
    As a rule, the diagnostic use of radiation does not increase risk of tumors. Exceptions are multiple fluoroscopies of pulmonary patients, for example, patients with tuberculosis. But even in this case, the cohort study of 64,172 tuberculosis patients in Canada has not detected any link between the risk of lung cancer and the dose. The report by A. Miller et al. [55], with high statistical reliability, shows a reduction, up

to 66% of the control group at a total dose of 15 cGy, of the relative risk of mortality from breast cancer in 31,710 women who repeatedly had X-ray screenings between 1930 and 1952.

13. Residents of areas with high natural radiation background [17].

Interesting information can be obtained by analyzing the variation in mortality or incidence of cancer associated with geographical variations of natural background radiation.

The residents of the mountainous state Colorado in the United States are exposed to radiation background of about 1.8 mSv/year, and the residents of Massachusetts, located on the Atlantic coast, only 1.02 mSv. If the linear no-threshold model is valid, it can be expected that the incidence of cancer in Colorado will be higher than in Massachusetts. If the dependence of cancer has a threshold that is higher than these values, one can expect that in both states, the risk is going to be the same. In 1999, studies showed that the probability of cancer in Massachusetts is 16% higher than in Colorado. This may be a manifestation of the radiation hormesis in the range of the dose and exposure conditions.

A similar result was obtained from more extensive comparisons of the population of the Rocky Mountains and coastal states. Results demonstrate undoubtedly that although the background radiation in mountains is higher, the frequency of cancers is lower. The same result is obtained in Canada. But a conscientious approach to data analysis needs to take into account all possible factors that could influence the result. Thus, a significant part of the population of Utah and of other mountain states make up the Mormons, who do not smoke or drink alcohol and coffee. Maybe this is the reason that the frequency of cancers in Mormons is half of that in their non-Mormon neighbors. On the other hand, the atmosphere-polluting industry is concentrated in the coastal states. With this in mind, the authors conclude that radiation, at least within the range of a few millisieverts a year, is not responsible for induction of cancer relative to other factors.

In India, epidemiological studies of five cities with different natural radioactive backgrounds showed a significant negative correlation between the level of natural radiation background in these cities and cancer mortality [56].

Similar studies conducted in China in areas with high radiation background revealed the same tendency to reduced mortality from cancer [57].

Let's add to all this the information about the inhabitants of the area with high man-made background. In Taiwan, about 10,000 residents of Taipei lived for 20 years in apartments whose walls had metal beams that contained ^{60}Co. The average radiation dose in these apartments was 340 mSv. And it turned out that instead of an increased incidence of cancer and leukemia, there was a sharp decline in mortality from cancer, which was only 3.6% of the average Taiwanese. Chronic exposure had led to residents' developing immunity from cancers. There was no information about developmental lag or developing tumors in children in irradiated apartments [23].

14. Miners at uranium mines. The main source of exposure is radon. For more details, see 4.6.
15. Naturally, it is necessary to draw special attention to the impact of radiation from the Chernobyl accident described in Section 6.5.4. It was the last cohort in the Russian edition written in 2011 and published in 2012. More exactly, in the case of Chernobyl, one needs to point out two cohorts: first are the liquidators, and the other the population of infected areas.
16. But since the first edition, new cohorts of irradiated populations appeared—the personnel, the team of workers, and the population of Fukushima in Japan, which is described in Section 6.5.5.

6.5.4 Chernobyl [58]

The Chernobyl disaster, as it is sometimes commonly called, was not the first accident at a nuclear power plant and is no longer the last. The history of nuclear accidents before Chernobyl at U.S. and Soviet nuclear facilities can be seen, for example, on the website [59].

But still, the Chernobyl disaster remains the largest accident during the existence of nuclear power and production of nuclear weapons. This disaster was one closest to our times that affected the lives of lots of people. Regarding the assessment of the accident's consequences, fierce debates are taking place and drastically different data get published.

In April 2011, the German division of the International Physicians for the Prevention of Nuclear War (IPPNW) and the Society for Radiation Protection (Gesellschaft für Strahlenschutz [GFS]) published the report "Health Effects of Chernobyl: 25 Years After the Accident" [60]. Here it is claimed that the International Atomic Energy Agency (IAEA) and the World Health Organization (WHO) understated the actual data on the expected cancer mortality, as compared with the data contained in the original research papers.

It is well known that the statements of some politicians or some biased journalists can show an obviously preconceived attitude to some research and even deliberate distortion of certain data. Let's apply the presumption of innocence to the scientists who publish research results in peer-reviewed journals. But it has to be understood that a noticeable spread of results can occur even with absolutely honest approach towards processing and analysis of observational data. The fact is that induction of cancer and cancer mortality, apart from the dose of radiation, also depends on a significant number of additional factors. The real impact of these factors and a competent accounting for, or, on the contrary, an underestimation of, their effects can significantly affect the final result. Examples of the influence on the result of studies of additional factors will be given below.

With the situation given, this brief review of the Chernobyl accident is going to contain the figures on radiation risk, morbidity, and mortality exclusively on the basis of scientific papers and data collected in the UNSCEAR and BEIR reports. As far as it is known, these organizations have never been accused of falsifying research materials.

The explosion of the Block 4 on the Chernobyl nuclear power plant occurred on 26 April 1986, at 01:23:47. Directly at the accident site, two plant employees were killed. Among the people who were carrying out emergency works, 134 cases of acute radiation sickness (ARS) were registered. By the end of 1986, 28 people died of ARS. Information on the early effects can be found in the report [61]. All the effects of radiation, which are going to be further discussed in this section, are stochastic, that is, long-term (Section 4.7.7).

In the liquidation of the accident aftermath, about 440,000 people were involved in 1986–1987. Later their number increased up to ~800,000. These people were called liquidators. A special medical register contains data as of 2008 on more than 500,000 liquidators. In the 30-km zone, about 200,000 people worked. Note that the number of liquidators included not only the workers, engineers, scientists, and military personnel who were actually engaged in the liquidation of consequences of the accident but also everyone who worked on the contaminated territories, including doctors, teachers, cooks, interpreters, and so on. From the 188 settlements in the affected areas, 115,000 people were evacuated in the spring and summer of 1986, and about 220,000 afterwards. On the Russian territory, 7500 settlements were contaminated, with a population of about 2.6 million people. By 2011, resulting from deactivation and natural decay, this number decreased up to 4000 (1.5 million inhabitants). The contaminated territories of the former Soviet Union with the total area of 150,000 km^2 still supporting by now up to five million people.

All the people exposed to radiation because of the accident were divided into the following groups:

a. liquidators; it is believed that the main stage of the accident's resulting liquidation lasted 3 years (1986–1989);
b. people evacuated from the town of Pripyat and from the 30-km zone around the destroyed reactor;
c. the population living in the affected area under strict control, the contamination by ^{137}Cs in this area exceeded 555 kBq/m^2 (15 Ci/km^2).
d. the population living in the area with ^{137}Cs contamination exceeding 37 kBq/m^2 (1 Ci/km^2).

Both the dose and the disease incidence for these groups are usually calculated and determined independently.

At the time of the accident, the reactor contained 192 tons of fuel and its decay products. This material was partly ejected by the explosion, partly melted and flowed down to the lower floors, and partly was discharged out into the atmosphere and groundwater, and is still being discharged as of now. According to a commonly accepted estimate, about 185 tons of nuclear fuel was left inside. Now its total activity is about 16 million curies. Most of it melted in the accident, mixed with melting construction fragments.

During the explosion and in the first 10 days, out of the active zone were ejected ~1760 PBq of ^{131}I ($T_{1/2}$ = 8.04 days), ~85 PBq of ^{137}Cs ($T_{1/2}$ = 30 years), ~10 PBq of ^{90}Sr ($T_{1/2}$ = 29.12 years), the isotopes of uranium, plutonium, iodine, cesium, strontium, and others. (Decimal prefix "peta" 10^{15} is rarely used, so here its meaning is going to be explained.) Let's point out that a lot of other radioactive nuclides also have been thrown out into the environment, but with short half-lives of the order of days or even hours. Also very long-lived nuclides emitted, such as ~0.013 PBq of ^{239}Pu with half-life $T_{1/2}$ ~ 24,000 years.

Furthermore, among the nuclear fission products were two inert gases, ~ 33 PBq of ^{85}Kr ($T_{1/2}$ = 10.72 years) and ~6500 PBq of ^{133}Xe ($T_{1/2}$ = 5.25 days). Krypton-85, because of its relatively large half-life, accumulates in the atmosphere, quickly gets stirred into the air of the northern hemisphere, and then, though much more slowly, penetrates into the southern hemisphere. For comparison, during the whole time of atmospheric nuclear weapons tests, 675,000 PBq of ^{131}I and 948 PBq of ^{137}Cs were discharged into the environment.

All radionuclides decay, and their concentration decreases with time, but one. This exception, ^{241}Am, appears in the environment after the decay of ^{241}Pu. Maximum activity of ^{241}Am will be reached, approximately, in 2058, and then it is going to slowly decline with a half-life of 432.2 years. At its peak, the activity of americium-241 will be about 1000 times lower than the activity of ^{137}Cs by that time.

The main radiation damage was created by the two radionuclides: the short-lived nuclide ^{131}I, which determined the impact on health during the first couple of months, and the gamma emitter ^{137}Cs, the main effect of which is currently ongoing.

As we know, iodine is concentrated in the thyroid gland, which naturally increases the dose. Conversely, cesium in the body behaves similarly to its chemical analog potassium, and therefore it gets quite uniformly distributed through the body, thereby reducing the effective dose.

In the first months after the accident, due to the lack, or inadequacy, of protection measures, the pollution of fresh milk by the nuclide ^{131}I caused significant radiation doses for the thyroid gland. At that time, children were subjected to the main radiation exposure. Subsequently, the activity of iodine dropped significantly and the entire population became exposed to irradiation mainly due to ^{137}Cs. Irradiation was both external and internal through contaminated food. However, because of protective measures, the resulting doses were relatively small. Thus, the average additional doses in the years 1986–2005 in the "contaminated" zone were equivalent to a dose from a session of computed tomography. The level of contamination by cesium-137 exceeds the limit of 37 kBk/km^2 on 45,000 km^2 on the territory of Europe (for reference, the area of, e.g., Germany is ~357,000 km^2 and Belarus 207,000 km^2) [58].

Evolution of various types of cancer was discussed in Section 4.4.7. It is shown in Figure 4.17. It is well known that acute exposure in a relatively short period of time (several years) results in development of leukemia. Maximum likelihood of disease is achieved about 6 years after exposure. To date, a maximum of leukemia should have been passed. Induction of new leukemia cases becomes almost zero about 30 years after exposure. The latency period of other cancer types is about 10 years, and the maximum is reached after 30–40 years.

The UNSCEAR report in 2008 provides information about the doses received by different groups, accounting for amendments relating to the improvement of treatment of dose recording techniques [58]. The average effective dose received by the liquidators during 1986–1990 mainly due to external radiation is about 120 mSv. Registered doses range from 10 mSv up to more than 1000 mSv, although approximately 85% of the workers received doses in the range 20–500 mSv. Estimates of uncertainty in individual doses range from

50% to five times. The experts noted that the doses received by military personnel are likely to have been understated.

Average additional effective dose received by the 6 million people living in the most contaminated area of Belarus, Ukraine and Russia during 1986–2005 from external exposure was about 9 mSv. For the 98 million people living in the three republics, who received 1.3 mSv, a third of the dose was received in 1986. These values are only slightly higher than the natural radiation background in the same period (~50 mSv). Over a 20-year period, about 70% of the population received an effective dose of less than 1 mSv and about 20% between 1 and 2 mSv. However, 150,000 people living in the affected area during the 20-year period received an effective dose of more than 50 mSv. For the 500 million inhabitants of other countries in Europe, the average effective dose is estimated at 0.3 mSv over this period.

Radiation doses received by the thyroid gland are noticeably higher. The average dose to the thyroid in 1986 was 490 mGy for the evacuees, 102 mGy for the population of contaminated area in the three republics, 16 mGy for the 98 million in Belarus, Ukraine, and 19 regions of Russia and 1.3 mGy for people in the remote countries of Europe. It is noted that preschoolers received a dose to the thyroid that was approximately two to four times greater than the average population.

Once again, we note that, currently, science cannot distinguish a tumor that developed as a result of irradiation from a tumor occurring for any other reason. Therefore, the link between the dose and morbidity is established by statistical methods. It is important to consider the impact on the disease probability of some factors such as age, gender, and genetic characteristics of the individual, lifestyle, and many other factors that may be relevant to the disease. To obtain reliable, sound epidemiological conclusions, it is necessary to have sufficient statistical support for the research.

Thyroid cancer is a rare disease in children. However, because the thyroid gland in children is quite an active organ, it has increased radiosensitivity. If a child has received a large dose of radiation to the thyroid and develops thyroid cancer several years after exposure, it is very likely that the cancer was caused by radiation. If thyroid cancer appears in an adult a few months after exposure, its relationship with exposure has very low probability.

An important factor that modifies the risk of radiation-induced thyroid cancer is iodine deficiency in human consumption. E. Cardis and colleagues found that in areas with low iodine content in soil, the risk is 3.1 times greater at 1 Gy, than in areas with higher levels of iodine. It is considered reliably proven that the use of potassium iodide as a dietary supplement significantly reduces the risk of radiation-induced thyroid cancer. It is clear that the underestimating of iodine deficiency, of this external, non–dose-related circumstance, significantly affects the risk assessment result.

Most of the researchers and experts point out that regardless of the exposure, the Chernobyl disaster has significantly changed the lives of millions of people, especially in Belarus, Russia, and Ukraine, who live in the most contaminated areas. Forced evacuation, limitations in usual activities, other countermeasures, and conflicting information about the possible consequences of the accident, have radically changed people's lives and led to psychological problems, worsened the quality of life, and could significantly affect health. Obviously, competent processing of epidemiologic data should not encourage to dismiss the unpleasant consequences at the expense of the accident to radiation phobia. On the other hand, the psychological aspects of the accident and postaccident conditions should be taken into account.

The increased incidence of thyroid cancer in children began to show in 1990–1991. This was pretty early for solid cancer, which is characterized by a much greater latency period, so this information initially caused doubt. However, further developments showed an obvious connection of these cancers with radiation. The number of these cancers is increasing steadily, and by 2005 no slowdown has been detected. According to the UNSCEAR report for 2008, among children who were younger than age 14 years at the time of the accident, for the period 1991–2005, a total of 5127 cases of thyroid cancer were revealed, and among those who were younger than age 18 years, 6848 cases. (In the UNSCEAR report for 2000, it was fewer than 1800 cases.) These numbers refer to Belarus and Ukraine wholly and to the four most affected areas of the Russian Federation. Background level of thyroid cancer in children under the age of 10 years is approximately two to four cases per year per million people.

For children born after 1986, no increase in frequency of thyroid cancer has been detected; the frequency is at the background level. There is no reliable information on thyroid cancer in exposed adults [58 p. 61, paragraph 70]. There is also no evidence of any increase in the frequency of any other solid cancers among the population of Ukraine and the Russian Federation [58, paragraph 77]. At the International Conference "Fifteen years after the Chernobyl disaster, Lessons Learned" in April 2001, it was noted that the mortality rate among liquidators and residents of contaminated territories does not exceed the statistical average data in the three countries. However, to assess such an exposure effect as mortality, the time that had passed after the disaster was still too short.

The report [17, p. 216–223] shows a table with an overview of 51 environmental studies on the population performed by domestic and foreign environmentalists. In most studies, an increased number of thyroid cancers are noted. In early works executed before 1995, no other types of cancer, including leukemia, were recorded, and only the subsequent papers reported of leukemia cases and some noncancerous diseases.

6.5.5 Fukushima Nuclear Accident

The most recent accident of the highest level of danger (7) began with the Great East Japan Earthquake of magnitude 9.0 at 2.46 p.m. local time on Friday March 11, 2011, that created the large tsunami. The earthquake was centered 130 km offshore the city of Sendai in Miyagi prefecture on the eastern coast of Honshu Island. The main blow fell on the Fukushima Daiichi Electric Power Plant located approximately 300 km north from Tokyo [62].

Eleven reactors at four nuclear power plants in the region were operating at the time and all shut down automatically when the quake hit. Subsequent inspection showed no significant damage to any reactor from the earthquake. The main problem arose almost an hour after the quake, when the entire site was flooded by the 15-m tsunami. Exactly at 3.42 p.m., three reactors, namely, the units 1, 2, and 3 at Fukushima, lost power. Because of a shutdown of the generators responsible for operation of cooling systems, three power blocks collapsed and lost the ability to maintain proper reactor cooling and water circulation functions. The temperature of the exposed fuel rose to some 2800 °C, so that the central part started to melt after a few hours, and by 16 hours after the scram (7 a.m. Saturday) most of it had fallen into the water at the bottom of the reactor pressure vessel. As a result, there was large-scale release of radioactive substances into the atmosphere, soil, and ocean.

Released radionuclides produced radiation rates about several hundreds of millisieverts per hour near the damaged units. Twenty kilometers northwest of the power plant on March 16, radiation levels of up to 0.33 mSv/hour was measured. The average radiation dose rate over the whole day in Ibaraki, between Fukushima and Tokyo, rose up to 0.1 mSv/hour but did not pose a health risk.

On March 17, IAEA radiation monitoring over 47 cities showed that levels of radiation in Tokyo had not risen. Although at some locations around 30 km from the Fukushima plant, the dose rates had risen significantly in the preceding 24 hours (in one location from 80 to 170 μSv/hour and in another from 26 to 95 μSv/hour).

Early in the morning the day of the accident, March 11, a nuclear emergency was declared and immediately the Fukushima Prefecture issued an evacuation order for people within 2 km of the plant. Later this day the radius of the banned area was raised several times and on Saturday 12, the evacuation zone was extended to 20 km. Out from the dangerous territories, approximately 156,000 people were displaced as of early 2013.

In contrast to Chernobyl, there were no seriously injured victims in Fukushima. On March 24, three workers were exposed to high levels of radiation, which caused two of them to require hospital treatment after radioactive water seeped through their protective clothing while working in unit 3. Nobody received doses that could produce acute radiation syndrome. So only stochastic effects, mainly cancer, are expected. But cancer has a long latency period. Too little time has passed since the accident and now it is very early to analyze the real health situation. Now one can only calculate the prognosis and make assessments of the possible future human casualties. Detailed information about risk assessment can be found in [63,64].

The radiation environment provides the basis to believe that for populations living in the most affected areas, there is a slightly increased risk of developing thyroid cancer, of leukemia, and solid cancers. It is known that 167 Fukushima plant workers received radiation doses that slightly elevate their risk of developing cancer. Reports of WHO and UNSCEAR indicate that outside the geographical areas most affected by radiation, the predicted risks remain low and no observable increases in cancer beyond the natural variation in baseline rates are anticipated. Estimated effective doses from the accident outside Japan are considered to be below (or far below) the dose levels regarded as very small by the international radiological protection community.

6.5.6 Conclusion to Radiation Hormesis

The list presented earlier does not cover all the possible cohorts that have received, or are receiving, doses of radiation that are increased, compared with natural background, but fall within the category "low." But the above list is enough to ensure that radiologists have significant material for analysis.

In these cohorts, in some cases there has been a sharp single irradiation with a high dose rate; in others, the cumulative dose was received over a long time at a low dose rate. In some cohorts, the irradiation was purely external; in others, it was both external and internal, and, say, the radiation associated with radon was only internal. In other words, the observational data collected includes a wide range of exposure conditions.

Summarizing, we can conclude the following. To date, we cannot say for sure that for all the exposure conditions, and all types of cancer and noncancer effects of exposure for different population groups, there is a threshold of stochastic effects and a range of favorable effects of radiation, that is, radiation hormesis. But the whole set of experimental (on bacteria, tissue culture, plants and animals) and observational data show that, apparently, both threshold and hormesis exist, and at least in many cases. The only area where the exposure to any doses exceeding the natural radiation background should be treated with great caution is fetal development. A developing organism has a particularly high sensitivity.

6.6 Danger and Safety of Low-Dose Radiation and Chemicals

All living things appeared, lived, and developed in conditions of natural background radiation. There is some evidence that radiation was one of the physical factors necessary for the emergence of life on Earth and creating the diversity of living organisms. Moreover, billions of years ago, at the time of the first organisms, background radiation was higher than it is nowadays by about three to five times. The whole history of the origin and evolution of life and of a man, in particular, proceeded under conditions of constant exposure to natural radiation background, of cosmic and terrestrial origin. There are still areas in the world (India, Brazil, China, Iran, France, the Caucasus, etc.), where the natural radiation background is 10 or more times higher than the Earth average. Numerous complex medical studies have not revealed any violations in the health of the local population as compared to regions with average levels of natural background radiation (see Sections 1.10 and 6.5.3). The experts note that in a large number of generations, the accumulation of genetic load that would not be compatible with the existence of plants and animals, including humans, has not happened [27,65].

None of the living organisms developed an organ sensitive to ionizing radiation, although organs were developed with stunning sensitivity and resolution to many other types of external influences—visible and thermal radiation, sound, smell, sound vibrations, and taste sensations. There is no doubt that if it was required in the process of evolution, that is, if sensitivity to ionizing radiation would have any adaptive significance, such an organ could have appeared.

But along with it, powerful protection systems have been developed in the human body by evolution. A system of recovery or removal of damaged molecules and cells (repair of DNA damages, membranes, regulation of intercellular relationships, apoptosis, etc.) has been developed and genetically entrenched.

The overwhelming majority of DNA damages in humans are spontaneous and they appear due to the thermal motion of the molecules, as well as from getting induced by free radicals formed during the metabolism of oxygen. Each mammalian cell undergoes about 70 million spontaneous DNA lesions per year [66]. Only the existence of a powerful and effective repair mechanism enables a living organism survive with so many damages [65].

Ionizing radiation also affects DNA, but much less frequently. At the present average natural radiation background of 2.4 mSv/year, radiation is responsible for no more than about five DNA damages in one cell per year. Human protection mechanisms protect us from harmful effects of radiation in a fairly wide range of doses—from lower than 1 mSv to higher than 280 mSv/year [30].

Exposure to low doses of radiation can be regarded as an irritating and thus stimulating factor and be compared, for example, with acupuncture, with wet cups, or mustard plasters. Charged particles like small needles bombard the body's cells and force them to defend themselves.

To illustrate the reasons why the linear no-threshold principle possibly doesn't work and small doses of radiation can be completely harmless or even beneficial, scientific literature provides many examples and analogies. Although the analogy is not at all the indication of proof, it shows the logical inconsistency of simple linear dependence. For example, Zbigniew Jaworowski indicates that the temperature 20 °C might seem dangerous, because it is obvious that 200 °C is enough to cause third-degree burns. And there is no reason to be afraid of a sip of cognac; it probably has a beneficial effect on human body, even though a gallon of alcohol is almost a lethal dose [65].

A good example of the incompetence of linear extrapolation is given by A. B. Koldobskiy and V. P. Nasonov [67]. If a car, having spent 10 L of petrol, can travel 100 km, one can be pretty sure that after spending 5 L it can travel 50 km. The question is how far the car will travel with 1 drop of gasoline. If the assessment is based on the drop volume of 0.1 cm^3 (10^{-4} L), then given the linear dependence on 1 drop, the car would make 1 m. However, an elementary common sense tells us that one drop entered into an empty tank won't move the car from the spot. The above example clearly shows that there is a threshold.

We now turn to the effects of low doses of chemicals. The overwhelming majority of all chemicals that a human being consumes with food are of natural origin. And among other products, there are also some natural pesticides, chemicals that plants produce to protect themselves from insects, fungi, and other pests. Each plant produces a whole set of such chemicals. There are quite a lot of various natural pesticides, but only 71 of them have been studied for carcinogenicity. And more than half of them have shown to be carcinogenic in animal experiments. Natural pesticides tested as carcinogens are certainly present in vegetables, fruits, herbs, and spices. It is believed that each American consumes an average of 1.5 g of natural pesticides per day.

Not only pesticides can be of concern. When cooking food, one produces about 2 g/person/day of burned products that contain many substances that show carcinogenic and mutagenic properties in tests. Some of the substances in roasted coffee often show carcinogenicity in tests.

But human diet includes not only natural products. A number of synthetic substances are also absorbed into the body, and this amount increases rapidly. It is often assumed that natural substances are part of the history of human evolution, and synthetic materials are recent products. So the mechanism developed in animals to suppress the toxicity of natural substances cannot cope with protection from synthetic substances. According to Ames, the leading expert in this field (the author of the bacterial test, which was named after him) and his coauthor Gould [68], this assumption is incorrect.

The fact is that multiple defense mechanisms are configured for general control of toxic substances rather than of some individual substances. If herbivores were protected against a specific set of toxins, they would have serious developmental problems in the transition to a new kind of food, if familiar foods became insufficient.

Ames and Gould [68], in particular, warn against the danger of overestimating such an acclaimed pesticide as DDT. Until now, there is no conclusive epidemiological evidence that DDT concentrations normally found in the environment or in human tissues, give a significant contribution to cancer. On the other hand, DDT is of considerable merit to mankind. DDT, the first synthetic pesticide, eliminated malaria in many places in the world, including the United States. It is very effective against many disease carriers such as mosquitoes, tsetse flies, lice, mites, and fleas. DDT is also fatal to many pests of crops and significantly increases productivity and reduces the cost of food, making fresh nutritious food more accessible to poor people. DDT is also slightly toxic to humans. In 1979, a report by the National Academy of Sciences of the United States said: "In less than two decades, DDT has prevented death due to malaria in 500 million people, which would otherwise be inevitable."

B. Ames wondered why so many substances in carcinogenicity tests give a positive result [68]. And he explains it by the effect of large doses as such, and not the properties of substances. Since, as a rule, the task of a test is to determine the value of a 50% probability of tumor occurrence, in standard tests for cancer, animals are given for a long time relatively large, almost toxic doses, close to the maximum tolerated dose (MTD). Toxic doses cause cell death and a compensating intensive division of surrounding cells. There is more and more evidence that cell division caused by large doses, but not by the substance, can lead to cancer in these tests [69–71]. Large doses may cause long-term tissue damage, cell death, and related continuous division of neighboring cells, which is a source of tumor development risk. Once a cell divides, there is each time some probability that a mutation will occur and, therefore, an increased frequency of cell division increases the risk of cancer. At low doses of synthetic substances, which people are usually subjected to, this increase of cell division frequency doesn't occur. Tissues damaged by large doses of chemicals give an active immune response, including activation of leukocytes in response to cell death. Activated leukocytes evolve mutagenic oxidants, including peroxynitrite, hydrogen peroxide, and hypochlorite. Therefore, small doses of chemicals to which a human being is exposed through contaminated water or residues (traces) of synthetic pesticides that do not affect the body's tissues can be generally safe or create minimal risk of cancer.

Note that the question is being seriously discussed, whether benefit for health from restricting the dietary energy value is related to the phenomenon of hormesis. The answer is yes, it's an obvious manifestation of hormesis [72]. The detrimental effect on health of excessive caloric intake is well known.

References

[1] Oberboum M, Camber J. Hormesis: dose dependent reverse effects of low and very low doses. In: Endler PC, Schulte J, editors. Ultra high dilution. Dordrecht: Kluwer Academic; 1994.

[2] Calabrese EJ, Baldwin LA. Tales of two similar hypotheses: the rise and fall of chemical and radiation hormesis. BELLE newsletter. 1999:8 (2) 47–66. http://www.belleonline.com/newsletters/volume8/vol8-3.html.

[3] Hormesis and Homeopathy. BELLE Newsletter. 2010;16 (1):1–3. http://www.belleonline.com/newsletters/volume16/vol16-1FINAL.pdf.

[4] Health Physics, v. 52 No. 5.(1987).

[5] Sanders CL. Radiation hormeses and the linear-no-threshold assumption. Springer; 2010; p. 217. ISBN: 978-3-642-03719-1.

[6] Mattson MP, Calabrese EJ, editors. Hormesis – a revolution in biology, toxicology and medicine. New York, NY: Humana Press; 2010. p. 208.

[7] Luckey TD. Hormesis with ionizing radiation. CRC Press; 1980.

[8] Luckey TD. Radiation hormesis. CRC Press; 1991.

[9] Cross F. Legal implications of hormesis. BELLE Newsletters 2001;9(2). http://www.belleonline.com/newsletters/volume9/vol9-2/n2v92.html.

[10] Cook R, Calabrese EJ. The importance of hormesis to public health. Environ. Health Perspect. 2006;114:1631–5.

[11] Calabrese EJ, Baldwin LA. Defining hormesis. Hum. Exp. Toxicol. 2002;21:91–7.

[12] Ivanovskiy YuA. Radiation hormesis. Are low doses of ionizing radiation beneficial. Vestnik FEO RAS 2006;6:86–91. [in Russian].

[13] Foekens JA, Sieuwerts AM, Stuurman-Smeets EMJ, et al. Pleiotropic actions of suramin on the proliferation of human breast cancer cells in vitro. Int. J. Cancer 1992;51:439–44.

[14] Kudryashov YuB. Radiation biophysics (ionizing radiation): a textbook. Moscow: Fizmatlit; 2004. p. 443 [in Russian].

[15] Eidus LKh, Eidus VL. Problems of the mechanism of radiation and chemical hormesis. Radiat. Biol. Radioecol. 2001;41(5):627–30. [in Russian].

[16] UNSCEAR 2000 Report. Annex G: biological effects at low radiation doses, p. 96 http://www.unscear.org/unscear/publications/2000_2.html.

[17] BEIR VII Phase 2 (2006) – Health risks from exposure to low levels of ionizing radiation. http://www.nap.edu/openbook.php?record_id(11340&page(R1.

[18] Burlakova EB, Konradov AA, Maltseva EL. Action of super low doses of biologically active substances and low intensity physical factors. Chem. Phys. 2003;22(2).

[19] Koterov A.N. Low doses of radiation: facts and myths. Book one. Basic concepts and genomic instability. Moscow: A.I. Burnasian FMBC of Russian FMBA; 2010, p. 283 [in Russian]. http://www.fmbcfmba.org/default.asp?id(6001111.

[20] Wilson R. Harvard University. Effects of ionizing radiation at low doses. http://users.physics.harvard.edu/~wilson/resource_letter.html#low.

[21] Vasilenko I. Ya., Vasilenko O.I. Radiation risk at low dose irradiation is negligible. http://nuclphys.sinp.msu.ru/ecology/risc.htm. [in Russian].

[22] Crawford M, Wilson R. Low-dose linearity: the rule or the exception? Hum. Ecol. Risk Assess. 1996;2(2):305–30.

[23] Buldakov LA, Kalistratova VS. Radiation impact on the body. The positive effects. Moscow: Informatom; 2005. p. 246 [in Russian].

[24] BELLE Newsletter. http://www.belleonline.com/newsletters.htm.

[25] Gilbert S.G. A small dose of toxicology. The health effects of common chemicals. 2nd ed. Healthy World Press; 2012, p. 280. http://www.toxipedia.org/download/attachments/9175184/A%20Small%20Dose%20of%20Toxicology%2C%202nd%20Edition.pdf?version(1&modificationDate(1331304565000&api(v2.

[26] Kuzin AM. The stimulatory effect of ionizing radiation on biological processes. Moscow, Atomizdat, 1977, 133 p. [in Russian].

[27] Kuzin AM. The idea of radiation hormesis in the atomic age. Moscow: Nauka; 1995. p. 158 [in Russian].

[28] Pierce DA, Vaeth M. The shape of the cancer mortality dose-response curve for atomic bomb survivors. Radiat. Res. 1991;126:36–42.

[29] UNSCEAR 1986 Report. Annex A: genetic effects of radiation; p. 27–164; Annex B: dose-response relationships for radiation induced cancer; p. 165–262. http://www.unscear.org/unscear/en/publications/1986.html.

[30] UNSCEAR 1993 Report. http://www.unscear.org/unscear/en/publications/1993.html.

[31] Sagan LA. What is hormesis and why haven't we heard about it before? Health Phys 1987;52(5):521–5.

[32] Planel H, Soleilhavoup JP, Tixador R, Caratero C. Influence of protection against natural irradiation in post-autogamous *Paramecium aurelia*. Comptes Rendus des Seances de La Societe de Biologie et de Ses Filiales 1968;162:990–5.

[33] Planel H, Soleillhavoup JP, Tixador R, et al. Influence on cell proliferation of background radiation or exposure to very low chronic gamma radiation. Health Phys. 1987;52(5):571–8.

[34] Luckey TD. Documented optimum and threshold for ionising radiation. Int. J. Nuclear Law 2007;1(4):378–409.

[35] Luckey TD. Radiation hormesis overview. RSO Magazine 2003;8(4):22–41.

[36] Luckey TD. Radiation hormesis: the good, the bad, and the ugly. Dose Response 2006;4(3):169–90.

[37] Calabrese EJ, McCarthy ME, Kenyon E. The occurrence of chemically induced hormesis. Health Phys. 1987;52(5):531–41.

[38] Calabrese EJ, Baldwin LA. The hormetic dose response model is more common than the threshold model in toxicology. Toxicol. Sci. 2003;71(2):246–50.

[39] Calabrese EJ, Stanek EJ III, Nascarella MA, Hoffmann GR. Hormesis predicts low-dose responses better than threshold models. Int. J. Toxicol. 2008;27:369–78.

[40] BELLE Newsletter. т.16, №1, 2010. Hormesis and homeopathy. http://www.belleonline.com/newsletters/volume16/vol16-1FINAL.pdf.

[41] Chaffey JT, Rosenthal DS, Moloney WD, Hellman S. Total body irradiation as treatment for lymphosarcoma. Int. J. Radiat. Oncol Biol. Phys. 1976;1:399–405.

[42] Jenner TJ, DeLara CM, O'Neill P, et al. Induction and rejoining of double strand breaks in V79-4 mammalian cells following α- and γ-irradiation. Int. J. Radiat. Biol. 1993;64:265–73.

[43] Goodhead DT. Initial events in the cellular effects of ionizing radiations: clustered damage in DNA. Int. J. Radiat. Biol. 1994;65:7–17.

[44] UNSCEAR 2006 Report. Annex A: epidemiological studies of radiation and cancer, paragraphs 93-98. http://www.unscear.org/unscear/en/publications/2006_1.html.

[45] BEIR VII Phase 2, 2006, 141–54. http://www.nap.edu/openbook.php?isbn=030909156X.

[46] Ozasa K, Shimizu Y, Suyama A, Kasagi F, Soda M, Eric J. Grant EJ, et al. Studies of the mortality of atomic bomb survivors, report 14, 1950-2003: an overview of cancer and noncancer deceases. Radiat. Res. 2012;177:229–43.

[47] Doss M, Egleston BL, Litwin S. Comments on "Studies of the mortality of atomic bomb survivors, report 14, 1950-2003: an overview of cancer and noncancer diseases. Radiat. Res. 177 (2012) 229-243". Radiat. Res. 2012;178(3):244–5.

[48] Doss M. Evidence supporting radiation hormesis in atomic bomb survivor cancer mortality data. Dose-Response 2012;10(4):584–92.

[49] The article in Wikipedia – Tsutomu Yamaguchi. http://en.wikipedia.org/wiki/Tsutomu_Yamaguchi.

[50] UNSCEAR 1994. Report. Annex B: adaptive responses to radiation in cells and organisms to the general assembly; p. 185–272. http://www.unscear.org/unscear/en/publications/1994.html.

[51] Gribbin MA, Weeks JL, Howe GR. Cancer mortality (1956-1985) among male employees of Atomic Energy Limited with respect to occupational exposure to low-linear-energy-transfer ionizing radiation. Radiat. Res. 1993;133(2):375–80.

[52] Abbat JD, Hamilton TR, Weeks JL. Epidemiological studies in three corporations covering the Canadian nuclear fuel cycle; Biological effects of low level radiation. Vienna: IAEA; 1983. p. 351.

[53] Darby SC, Kendall GM, Fell TP, et al. A summary of mortality and incidence of cancer in men from the United Kingdom who participated in the United Kingdom atmospheric weapons tests and experimental programs. Br. Med. J. 1988;296:332–40.

[54] Raman E, Dulberg CS, Spasoff RA. Mortality among Canadian military personnel exposed to low-dose radiation. Canad. Med. Assoc. J. 1987;136:1951–5.

[55] Miller AB, Howe GR, Sherman GJ, et al. Mortality from breast cancer after irradiation during fluoroscopic examination in patients being treated for tuberculosis. N. Eng. J. Med. 1989;321:1285–9.

[56] Nambi KSV, Soman SD. Environmental radiation and cancer in India. Health Phys. 1987;52(5):653–7.

[57] Wei L, Zha Y, Tao Z, He W, Chen D, Yaan Y. Epidemiological investigation of radiological effects in high background radiation areas of Yangjiang China. J. Radiat. Res. 1990;31(1):119–36.

[58] UNSCEAR 2008 Report, v.II. Annex D: health effects due to radiation from the Chernobyl accident; p. 45–219. http://www.unscear.org/unscear/en/publications/2008_2.html.

[59] Chernobyl, Pripyat, Chernobyl AEP and exclusion zone. http://chornobyl.in.ua/istoria-avariy.html [in Russian].

[60] Health Effects of Chernobyl. 25 years after the reactor catastrophe. 2011, IPPNW (International Physicians for the Prevention of Nuclear War) and GFS (Gesellshaft für Strahlenschutz) Report. http://www.ippnw.org/pdf/chernobyl-health-effects-2011-english.pdf.

[61] UNCSCEAR 1988 Report. Annex G: early effects in man of high dose radiation. http://www.unscear.org/unscear/en/publications/1988.html.

[62] World Nuclear Association. Fukushima Accident (August 2014). http://www.world-nuclear.org/info/safety-and-security/safety-of-plants/fukushima-accident/.

[63] Radiation effects from the Fukushima Daiichi nuclear disaster [article in Wikipedia]. http://en.wikipedia.org/wiki/Radiation_effects_from_the_Fukushima_Daiichi_nuclear_disaster.

[64] World Health Organization. Health risk assessment from the nuclear accident after the 2011 Great East Japan Earthquake and Tsunami based on a preliminary dose estimation. WHO, 2013, p. 172. http://apps.who.int/iris/bitstream/10665/78218/1/9789241505130_eng.pdf?ua=1.

[65] Jaworowski Z. Radiation Risk and Ethics. Physics Today 1999;52(9):24–9.

[66] Billen D. Spontaneous DNA damage and its significance for the "negligible dose" controversy in radiation protection. BELLE Newsletter 1984;3(1):8. http://www.belleonline.com/newsletters/volume3/vol3-1.html.

[67] Koldobskiy AB, Nasonov VP. Around atomic energy: truth and fiction [in Russian]. Moscow: MEPhI; 2002. p. 116.

[68] Ames BN, Gold LS. Paracelsus to parascience: the environmental cancer distraction. Mutat. Res. 2000;447:3–13.

[69] Heddle JA. The role of proliferation in the origin of mutations in mammalian cells. Drug Metab. Rev. 1998;30:327–38.

[70] Gold LS, Slonc TH, Ames BN. What do animal cancer tests tell us about human cancer risk? Overview of analyses of the Carcinogenic Potency Database. Drug Metab. Rev. 1998;30:359–404.

[71] Canningham ML. Role of increased DNA replication in the carcinogenic risk of nonmutagenic chemical carcinogens. Mutat. Res. 1996;365:59–69.

[72] Turturro A, Hass BS, Hart RW. Does caloric restriction induce hormesis? BELLE Newsletter 2000;8(3). http://www.belleonline.com/newsletters/volume8/vol8-3/n2v83.html.

7 The Synergic Effect of Radiation and Chemical Agents

The living organism can be exposed to the one single factor exclusively in a pure, thoroughly organized experiment. However, in real-world circumstances and, in particular, in epidemiological studies, the organism is exposed to several factors. The effect of all these factors can be cumulative. This combined effect is called "additivity." But on several occasions, some factors can either intensify or reduce the effect of the others. If the cumulative effect of the two or more factors significantly exceeds the effect of each separate component, such a phenomenon is called "synergism." Some aspects of synergism are going to be discussed in this chapter. Also of interest is a state where one of the factors, that is, a certain agent, reduces the effect of another factor, for example, radiation. Thus, the effect of "antagonism" emerges. This problem is examined in Chapter 8.

While discussing the deviation from "additivity," one has to consider the two types of combined effects. In the first case, both factors, radiation and a chemical agent, independently create a poisonous effect. In the second case, the chemical agent, though harmless in itself, intensifies the radiation effect.

In the case of linear connection dose–response the concepts "additivity" and "synergism" are quite evident. But for a nonlinear connection, which is often observed in practice, some precise definition is required. An example of the dose–response relationship with a growing derivative is shown in Figure 7.1. If in the dose–response relationship, the form of a functional dependence after the initial interaction of factors does not change, this interaction of factors is called "isoadditivity." If the two factors act quite independently, the result is "heteroadditivity." The whole range of possible effect values between the two extremes is the range of additivity. It can be clearly seen from the diagram how, depending on the dose, the effect changes with occurrence of synergism and antagonism.

It has to be noted that multiple analysis of common influence of various agents on the biosystem is a part of an extensive division of toxicology [1].

The detailed review of synergic effects and the analysis of possible mechanisms of this phenomenon can be found in the UNSCEAR report for 2000 [1].

Generally, with the complete account of synergism as a phenomenon, the combinations of various physical, chemical, biological, and sometimes psychological factors are considered. In the UNSCEAR report [1], in particular, one can find the information on combined action of ionizing and ultraviolet radiation, of ionizing radiation and radio waves, or high temperature, or dust, or a stress situation of a space flight, etc. But here we are going to analyze only the interaction of radiation and chemical agents.

As in the environment there exist a great number of chemical substances, which can have a real influence on radiation effects, it is essential to present the classification of these

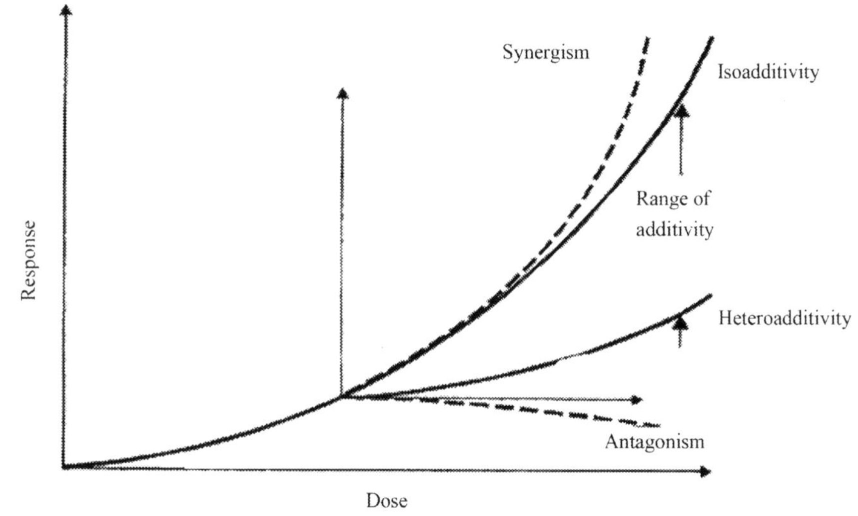

FIGURE 7.1 Interaction of two factors with non-linear dependence dose–response. Based on [1].

substances, which is rather difficult to make, as there is a great scale of options for the actions of various substances. But roughly all these substances could be subdivided into two groups according to the way of their interaction with DNA. The first group is the substances, which can, directly or by free radicals, produce defects in DNA, or so-called genotoxic substances. To the other group belong substances that are actually nontoxic but are able to influence the regulating and controlling systems of cells and organs.

One of the evident manifestations of synergism is the so-called oxygen effect (see Section 4.4.2). Oxygen is known for its genotoxic activity. It was already noted in Section 4.4.2 that the presence of oxygen leads to the more severe radiation damages than can occur from radiation exposure in its absence. The highest known values of COG (the Coefficient of Oxygen Gain) are of the order of 3 and are gained with the normal oxygen content in the atmosphere (~21% = 159 mm Hg). Thus comes the conclusion that if the substances that can temporally reduce the concentration of oxygen in the body tissues are to be injected into the cells, that can make the whole organism more tolerant to the radiation effect. This method together with the other ones that protect the human body from radiation is going to be fully considered in Chapter 8.

7.1 Smoking

The synergy of radiation and smoking effects are quite evident. A large number of meticulously conducted studies revealed a considerable increase with the smokers of a radiation-induced risk of cancer. Section 3.1, where the methods of risk calculations are considered, offers the example of occurrence of lung cancer with the smokers. This example is based on real research data. Here the relative risk of cancer for the smokers equals 15, which means that the risk of cancer for smokers within the cohorts under study was 15 times greater than for nonsmokers.

The lung cancer takes a significant position amid the other types of cancer, while the lungs are the way of entering of many attacking factors into the human body. Thus, the considerable doses of radon, to which the workers in the uranium mines were exposed, cause lung cancer. In some of the early studies conducted in the United States, it was found that white miners were subjected to lung cancer more often than the aboriginals, the American Indians, were. At first, there were suggestions toward ethnic and racial difference as far as predisposition to cancer was concerned. But further monitoring of the miners showed that there were much fewer smokers among the Indians, which was crucial in defining the difference in the likelihood of disease [2].

Nevertheless, detailed analysis of particular substances that are the constituents of tobacco smoke is no easy matter, because tobacco smoke is a compound mixture of genotoxic and nongenotoxic substances. Besides, tobacco smoke contains some nuclides and long-lived decay products of Radon ^{210}Po ($T_{1/2} = 138.4$ days) and ^{210}Pb ($T_{1/2} = 22.2$ years).

The specialists point out that although it has been already identified about 400 chemical components in tobacco smoke, but probably, there are some as yet unidentified substances that are short-lived or present in very small concentrations though super-active.

7.2 The Diet

Essentially, a certain interaction with radiation and, among other factors, synergism can also be observed in some foodstuffs. It is evident that a human being gets from food all the necessary components needed for building and functioning of the cells. The food has all kinds of substances that can produce superadditive as well as subadditive effects. Synergic action can be expected from food with a rather small amount of radical converters or coenzymes that are required for the effective repair of DNA and other cell components. Experiments on animals have shown that effects of intensifying the induction of tumors can be observed with the increased consumption of such substances as riboflavin, ethanol, or marijuana. The effects of reducing the induction of tumors can be observed with the low-calorie diet; with consumption of vitamins A, C, K, and E; selenium; and some other food components.

7.3 Problems of Radiation Therapy

An earlier section emphasized the possibility of intensification of the radiation effect along with the simultaneous action of some other chemical agent. This intensification was considered an undesirable factor that makes radiation more dangerous. But there are a few quite real practical problems when the intensification of cell radiosensitivity is desirable, such as for tumor radiotherapy.

For effective tumor destruction under the action of radiation, large doses are required. But the possibilities for increasing doses are limited, because after a certain limit, the other tissues of a patient start to be damaged. In order to solve this problem, a large number of different techniques were developed and implemented in practice. Some allow to intensify the radiation effect on the tumor without increasing the dose. Some are meant to reduce the radiation effect on healthy tissues without lowering the needed therapeutic doses. The phenomenon of synergism and, along with it, ways to intensify the radiation effect on tissues with the help of chemical modifiers without intensifying the dose are discussed here.

Let's discuss the possibilities of combining radio- and chemotherapy.

In fact, the situation when the breakage of one DNA bond, related to the formation of an adduct by a carcinogenic substance, can take place simultaneously with a breakage in the other bond, caused by a charged particle. This variant of the combination of factors was suggested to explain the interaction between cisplatin and radiation.

Many substances are known to inhibit the reparation of radiation damages. These can be antitumoral antibiotics, such as dactinomycin and doxorubicin, antimetabolites, for example, hydroxyurea, cytarabine, and some other substances.

One more possibility to increase the radiation effect is the activation of apoptosis by certain substances, that is, the stimulation of the cell death process with a simultaneous radiation exposure.

It was already pointed out that the increase of cell radiosensitivity can be achieved with the help of an oxygen effect. Usually, for this purpose, special sensitizers or electron-seeking compounds are used, whose action is similar to that of oxygen but they can better penetrate deep into the tumor.

Although much in the phenomenon of synergism is still unclear, one of the most famous Russian radiobiologists, A. M. Kuzin, pointed out that all living things on Earth are subject to exposure to many physical and chemical factors that operate concurrently with radiation [3]. Which are expected to be the outcomes of simultaneous action of ionizing radiation and of radio waves of different ranges, and of ultraviolet and infra-red radiation? What is going to be the influence of radiation in hot climate on the equator and at cold temperatures in the Far North? Is synergism going to reveal itself with a mutagenic effect of radiation together with the impact of chemical mutagenes, which day after day pollute our environment? What is going to be the effect of the action of small radiation doses for the big industrial cities, where the air is polluted by car exhausts, nitrogen oxides, and sulfur from chemical plants? All these problems require an intensive study.

References

[1] UNSCEAR 2000 Report. Sources and Effects of Ionizing Radiation. Volume II: EFFECTS, Annex H: Combined effects of radiation and other agents, p.179-295 - http://www.unscear.org/docs/reports/annexh.pdf.

[2] Whittemore AS, McMillan A. Lung cancer mortality among U.S. uranium miners: a reappraisal. J. Natl. Canc. Inst. 1983;71:489–99.

[3] Kuzin AM. Invisible rays around us. Nauka, Moscow, 1980, 151 p. – in Russian.

8

The Methods of Pharmacological Defense: Antidotes, Antimutagens, Anticarcinogens, and Radioprotectors

8.1 Antidotes

"Useless!" said Athos, "useless! For the poison which SHE pours there is no antidote" [1]. The lovers of historical novels of Alexandre Dumas, père, could have surely surmised that SHE is the treacherous Milady who has just poisoned the gentle Constance Bonacieux, d'Artagnan's lover. Poisons as well as antidotes are frequently present in historical novels. They were known in ancient times too, but it was only in the Middles Ages that they became so popular. However, the heyday of poisoning is the later medieval, that is to say, the achievement of "civilization." From the earliest times, the poisoners as real people and as fictitious characters, both male and female, were infamous though rather popular haunters of different stories, novels, and tales.

From time immemorial, there was an idea that if nature has created a poison, it also could have produced an antidote for it. One only has to be able to find it.

Unfortunately, there are no universal antidotes, though there exists mixed-activity substances. The antidotes have several options of action. First, its adsorption, which activated carbon offers. But this option is effective only if used at the early stage of infection, before the toxic agents are absorbed into the bloodstream. Second, antidotes can enter into a chemical reaction with a poison and thus are able to turn it into a harmless or a less-toxic and a quite soluble, low-molecular-weight substance, which could be easily freed from the body. And then there are antidotes that do not affect the toxic substances directly but can enter with them into a competition for the influence on reactive systems of the body. That's what the use of oxygen as an antidote to carbon monoxide poisoning is based on.

Except from obvious poisons, some medical products may also cause the toxic effects, for apart from curative action they could bring about a diversity of side effects from the quite simple, unpleasant, though not dangerous, right up to fatal ones that could lead to the death of a patient. As usual, the unfortunate consequences may occur from an overdose of some medicine, but they can also happen sometimes from an individual's sensitivity toward certain chemicals that could provoke possible acute allergic reactions, or laryngeal edema, or anaphylactic shock or some other severe effects.

Let's point out some relatively well-known substances and remedies that may function as antidotes. They are ascorbic acid and glucose – the substances with a relatively wide range of activity. Ethanol can be used at poisoning with methyl alcohol or ethylene glycol. But obviously it is only a medical specialist who might be able to determine the right medication.

The importance of defining the type of a poison in order to ensure the right antidote can be illustrated by the gas poisoning of both the terrorists and the people who came to the show "Nord-Ost" at the Dubrovka Theatre Center in Moscow in October 2002. The medical professionals were not prepared to determine the antidote for an unknown toxic substance, and that resulted in the death of a large number of people, including children.

8.2 Methods of Chemical Defense From Carcinogens

As far as there is no way to avoid the exposure to carcinogens, it is essential somehow to initiate the inhibition of carcinogenesis. One can try and use some chemical compounds, anti-mutagens, and anticarcinogens as preventive measures and in order to slow down the induction of mutation.

The phenomenon of antimutagenesis has been known already for a long time, as it was discovered in 1952 by A. Novik and L. Szilard and since then has demonstrated convincing experimental proof [2]. The situation with carcinogenesis is more complicated as this phenomenon has not yet been completely explained [2]. There are, though, some carcinogenic inhibitors of various types.

One of the ways of defense from genotoxicants is picking up the carcinogens outside the cell and thus preventing them to reach their target cells. To some extent, its effect is similar to an antidote's action. Dietary fibers that are present in fruit, vegetables, and some cereals by their sorbent ability promote the removal of a toxic substance from the body. Ascorbic acid as well as some other chemical compounds such as tannin acid and tocopherol are known to be able to inhibit the endogenic formation of genotoxicants. The independent epidemiologic research on a large group of volunteers has proved the positive role of ascorbic acid, fresh fruit, and vegetables in really decreasing the risk of cancer. A number of strong antimutagenic compounds have been found in apples, ginger, green tea, cabbage, pineapple, green pepper, and mint leaves [2].

Tumor development caused by carcinogens of a wide spectrum can be effectively retarded by the substances that act like antioxidants. One should note, though, that some antioxidants that can be active at the stage of initiation may have the opposite effect on already initiated cells, for example, they could exhibit a promoting effect.

Some other group of substances acts at the intracellular level, excluding the initiation of carcinogenesis. It includes ferments or substances favorable to enzyme induction that are able to change the speed and the metabolic ways of genotoxicity. Nevertheless, it is essential to note that the effect of these substances is rather specific and selective; that is, certain modifiers inhibit carcinogenesis caused by a small amount of carcinogens. Besides, these substances combined with other toxicants can either show no effect on carcinogenesis or even stimulate it. In [2] are shown many instances of such anticarcinogenic behavior.

The probability of inhibition in the initial stages of carcinogenesis is also determined by certain influence on reparation systems. The induction of enzymes that perform this reparation and also the activation of genes directly controlling the whole process could be both possible in this case. It's extremely interesting, especially in the context of this book, that activation of reparation systems in a defective DNA can be witnessed under exposure to small doses of ultraviolet and gamma rays that activate DNA-polymerase.

Thus, the data we have available now allow us for the time being to develop and apply the techniques toward the inhibition of carcinogenic initiation only experimentally or for the narrow groups of industrial workers with high cancer risk.

8.3 Radioprotectors

With the development of nuclear weapons during World War II, the possible doses of human irradiation dramatically increased, and the first victims of radiation sickness (see Section 4.5) appeared. This has stimulated the search for protective measures. The first chemical compounds that can reduce the damaging action of ionizing radiation were discovered more than 70 years ago. In 1942, it was discovered that it is possible to reduce the damaging action of ionizing radiation on enzymes by adding to the solution a number of substances that capture radicals. In 1948, a similar protective effect was achieved in experiments with bacteriophages. In 1949, the ability of a number of substances to protect mammals from radiation damage was proved. Experiments showed that the injection of cysteine 10 minutes before irradiation protected rats from a fatal radiation dose. Then the similar effect of cyanide on mice was discovered. The considerable work of many scientific laboratories all over the world lead to the establishment of a new trend called "modification of radiosensitivity in biological objects." That's how radioprotectors made their appearance.

It was found out that radioprotectors can have a protective effect if only they are injected into the body 10–30 minutes before the radiation. A radioprotector is useless if injected after radiation. Thus, one can conclude that radioprotectors can protect only during a short time of exposure. Nevertheless, the main bulk of research of their protective effects was carried out with considerable doses of radiation.

But recently, there have been news reports of some remedies that are able to protect a living organism even if it is injected after irradiation [3]. The mentioned study informs that the addition of antioxidant Tempol to the mice food allowed to reduce the risk of cancer and to increase the survival rate after considerable, though not fatal, whole-body radiation dose.

Among the first radioprotectors were the substances whose molecules contain a thiolic (–SH) and/or an aminic (–NH_2) group. Then the choice of radioprotectors was widened. The group of modern effective radioprotectors includes mercaptoethylamine – MEA, cisteamine, serotonin and some other substances. The main governing idea in the development of radioprotectors is the reduction of oxygen content in the tissues of animals, inactivation of free radicals and inhibition of free radical oxidation processes.

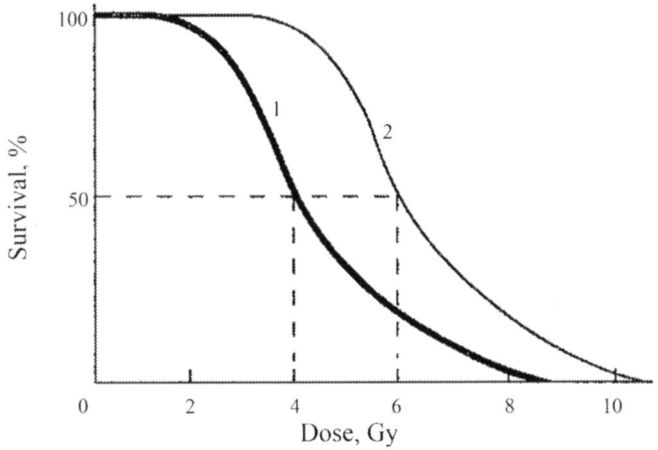

FIGURE 8.1 The dependence of survival of mice on X-ray dose.

The indication of protective effectiveness of a radioprotector is a value of the so-called dose reduction factor (DRF). It shows how much the equal effective dose increases with the application of certain preparations. Figure 8.1 shows the relative curves of the survival rate of the irradiated specimen with the use and without radioprotector. In this example, DRF = 1.5. DRF with the most effective radioprotectors is 1.8–2.5. The high toxicity of these preparations prevents to increase the value of this factor.

Georgy Gamov, one of the famous physicists (and among biologists, he is the author of the genetic code) of the twentieth century, thus shares his opinion on chemical protection:

> *"Take a good sip from this bottle (of an excellent Scotch Whisky) a couple of minutes before you are exposed to radiation, and your chances for survival after the atomic explosion would increase substantially." "Are you kidding me?" Mr Tompkins said incredulously. . . . "Not at all, it is a scientifically established fact. But, I repeat, whiskey must be taken before you will be exposed to radiation. Reception after would not mitigate the harmful effects of radiation." [4]*

The important field of application of radioprotective remedies is that of a radiation therapy. It was already noted in Section 7.3.

The above-discussed possibilities of chemical protection can be referred to the effects of large and even fatal radiation doses. But to such effects a human being is exposed in exceptional cases; besides, it is not at all evident beforehand that one would get a considerable radiation dose so that one could in advance take advantage of a protective remedy. Far greater numbers of people could be exposed to smaller doses of chronic radiation. That can be seen in an example presented in the Section 6.5.3 when the cohorts were exposed to radiation. That's why it is so essential and equally interesting to search for protective ways from the persistent effect of small radiation doses. Numerous research studies in many countries, Russia included, have revealed that for that purpose it is quite reasonable to make use of the biologically active substances of natural origin. Owing to the absence of toxicity as well as to the good tolerance, these biologically active substances can be used as food supplements. As expected, the application of these supplements can increase the overall nonspecific resistance of the body by stimulation of the protective antioxidant body reserves.

The excretion of radionuclides from the organism is one of the special methods of chemical protection. The reason is that apart from the external radiation, the organism can be exposed to the internal radiation by radionuclides that enter the body through breathing, are ingested with food and drinks, and stay put in organs and tissues. Radionuclides located in the human body are called incorporated. The further inner radiation can be stopped by excretion of radionuclides as soon as possible. The fist help can come from such ordinary and, so to speak, mechanical sources as saunas, which cause excessive sweating as well as gastric and intestinal lavage. Apart from mechanical methods, there can also be help from such remedies as laxatives and diuretics, as well as preparations of stable isotopes that substitute their radioactive analogs and sorbents.

References

[1] Alexandre Dumas, père. Three musketeers. -http://www.gutenberg.org/cache/epub/1257/pg1257.pdf, page dxxxix.

[2] Khudoley VV. Carcinogens: Characteristics, Patterns, Mechanisms of action. Publ. Inst. Of Chemistry, St. Petersburg, 1999, ch. 7, p. 249-268 – in Russian.

[3] Mitchell JB, Anver MR, Sowers AL, Rosenberg PS, Figueroa M, Thelford A, et al. The antioxidant tempol reduces carcinogenesis and enhances survival in mice when administered after nonlethal total body radiation. Cancer Res. 2012;72(18):4846–55.

[4] Gamov G, Yčas M. Mr Tompkins inside himself: Adventures in the New Biology. New York, NY: Viking Press; 1967.

9 The Regulation of Radiation and Chemical Safety

9.1 The Regulation of Radiation Safety

The official date of X-ray discovery is November 1895, when once towards the night unexpectedly Wilhelm Konrad Röntgen in his laboratory at the Würzbirg University discovered the scintillation of a barium platinum cyanide screen induced by a cathode tube that was switched on, but completely covered with black paper. The first original paper by W. K. Röntgen, "On a New Kind of Rays" (Über eine neue Art von Strahlen), was published on December 28, 1895.

Science historians attribute the discovery of radioactivity to November 8, 1895, when Henri Becquerel, also just by accident, discovered, that a photographic plate folded in light-tight paper was being lighted by some rays outgoing from a sample of the doubled salt of uranil potassium. On November 23, 1896, Becquerel reported the property of uranium to radiate.

In the scale of a century, these two discoveries took place virtually at the same time.

In 1897, Marie Skłodowska-Curie together with her husband Pierre Curie joined the research of radioactivity. Having checked for radioactivity almost all known chemical elements by then, Marie Curie found that, apart from uranic ores, thorium salts may exhibit radioactivity. And by the middle of 1898, she and her husband Pierre discovered new chemical elements in a uranic mineral (a tar spooler): first radium and a little later, polonium. Since then, X-rays and radium radiation have been studied and used in parallel.

Soon after that, yet before the beginning of the twentieth century, the two most important peculiarities of newly discovered radiations were revealed.

First, it was found to be possible to use the penetrating rays for medical diagnostics. The immediately appreciated practical value of both discoveries was approved by the Nobel Committee, with the prize first awarded to Röntgen in 1901 (that was the first Nobel Prize ever given) and then in 1903 to A. Becquerel and to the Curie couple.

Then, also rather soon, it was made clear that these newly discovered radiations are dangerous. Thus, in 1895, W. Grubbe, one of Röntgen's assistant, operating with the X-rays got a severe burn on both of his hands. The skin of Becquerel, the discoverer of radioactivity, was badly burnt by radium rays too. T. A. Edison, the famous American inventor, who soon after W. K. Röntgen's discovery was working hard with the cathode tubes, found out that the many-hour activity with cathode tubes cause dermatitis, pain in the eyes, and headaches. The Curie couple too experienced agonizing, long rankling burns. The Russian scientists E.S. London and S. V. Goldberg also from self-experience confirmed that radium rays cause severe burns. They also found out in 1903 that radium rays are destructive for cancer cells. Thus, it was made clear that these two radiations can be used for curing tumors and other unwanted formations.

Both physicists, who studied the radioactive rays, and physicians, who used the results in practice, during the initial years experimented with them without any precaution or protection. Examining their patients with X-rays, the physicians themselves each day got a certain dose of radiation. The hidden damage of these rays was accumulated day by day. After 10–15 years of this common practice, a malignant tumor came in a mass attack against the roentgenologists. In a couple of years, almost all enthusiasts of this medical activity died.

In 1936 in Hamburg in front of St. Georg Krankenhaus (Hospital), the monument "To the Victims of Radium and X-Rays" was erected. On it were engraved the names of 110 scientists and engineers who fell victim to the first experiments with X-rays. Since then, this list has been continuously replenished so that very soon it had to be surrounded with new memorial stones. Up to 1959, there were already 360 names.

By and by, it became evident that certain measures had to be taken, so in 1913 the German Radiological Society suggested the first steps to protect radiologists from radiation damage.

In 1915 and then in 1921, the X-Ray and Radium Protection Committee of the British Royal Society of Medicine published a report on possible risks and protective measures to defend when operating X-rays and radium. In 1922, G. Pfahler addressing the American Radium Society suggested a similar list of measures to protect the employees operating with radium and X-rays. All those measures were not at all quite distinct and efficient, while physicists and radiologists have not, up to then, quite learnt to carry out reliable measurements of radiation doses. They have not as yet defined the concrete biological effects produced under the effect of certain doses.

Probably, the first important step in the needed direction was the initiation of the tolerant dose conception that was formulated in 1925 by A. Mutscheller. While the most evident outcome of radiation effect on the organism were skin burns, A. Mutscheller suggested to take as acceptable the monthly doses of less than 1/100 from the threshold of skin erythema occurrence. Around the same time, but completely independent, R. Sievert suggested the yearly dose of 1/10 from the threshold of skin erythema, and a group of British scientists suggested the daily dose of 0.00028 part of the value. It is easy to calculate that all those three independently suggested values result in quite the same amount [1].

In 1928, the International Commission on Radiological Units and Measurements (ICRU) was the first to standardize the definition of a measure unit, a roentgen. Then the International Committee on X-ray and Radium Protection (ICXRP) was organized in order to establish the safety standards for the operation with X-rays and radium. It was the first division of the International Commission on Radiological Protection (ICRP).

In 1934 the Advisory Committee on X-ray and Radium Protection (ACXRP) established the standard that was for the first time based on measurable units. This standard was equal to 0.1 R/day and approximately 25 R/year (the standards, suggested before by A. Mutscheller, but reduced by half). Soon after that, the ICXRP published a report in which a boundary value of 0.2 R/day was suggested.

With the extension of research and with accumulation of new information on radiation effects, it was getting clearer that ionizing radiation is dangerous not only for the direct injury of organs and tissues but also for the probable genetic consequences. Based on this, there was a suggested fivefold decrease of the tolerable radiation limit, up to 0.02 R/day (~ 5 R/year) [2].

After the fission of uranium was discovered and the Second World War began, and especially with the activity in creating the atomic weapons, the research of biological effects of ionizing radiation became more intense. It was found that there were many different nuclides with different properties that carried different levels of dangers for the organism. It became clear that not only the external but also internal radiation is possible; besides they have different dose distribution and perform different effects. It caused to bring into use the extreme values of internal content of radioactive contamination as well as of radionuclide concentration in the air, in water, and in food products. There appeared the first instances of radiation sickness (see Section 4.5). The physicians before were dealing mainly with local lesions or with effects that we now call stochastic effects.

After the emergence of atomic stations came a large number of companies producing and processing nuclear fuel and nuclear weapons, as well as developing, producing, and operating the radionuclide devices, and the people who could actually be exposed to the radiation were subdivided into two groups, the employees and the population. The former are the people who work directly with the sources of radioactivity. They are healthy adults who before their employment are medically examined and, then, in the course of their service, the status of their health is periodically checked. The latter group comprises all the rest, including those who are less protected from radiation effects, that is, children, pregnant women, elderly, and sick people. It is evident that the boundary values for these groups might be different.

In 1960, at the 11th General Conference on Weights and Measures accepted the standard that was called "The International System of Units" (in French: "Système Inrenational d'Unités" [SI]). After that, it started to take roots in different countries. According to the new system, the ICRP recommended making use of the units of the absorbed, equivalent, and effective radiation doses, the gray and the sievert (see Section 1.8.1), instead of the units of roentgen, rad, roentgen equivalent physical (rep), and roentgen equivalent man (rem).

From 1941 and up to the present day, the safety standards have been revised several times. It is essential to note that all quotas were determined from the measuring data of large doses with the use of the linear no-threshold hypothesis.

In order to exclude the nonstochastic effects, no part of the human body should get more than 0.5 Sv/year except crystalline lens, the permissible dose for which is limited by the value of 0.15 Sv. In order to reduce the stochastic effects, the whole body irradiation dose must be limited by the value of 0.05 Sv/year.

The standards of radiation safety determine the values of effective dose: for the employees, it is an average of 20 mSv/year for any successive 5 years, but not exceeding 50 mSv/year. For the rest of the people, an average of 1 mSv/year for any successive 5 years but not exceeding 50 mSv/year.

The effective dose for the employees must not exceed 1000 mSv for the period of labor activity (50 years), and for the rest of population it must not exceed 70 mSv for the life span (70 years).

The standards of radiation safety regulate in detail various options of irradiation with sealed and unsealed sources – natural, medical, by radiation accident, the irradiation of different groups of population, of various organs. It might be well to note that the permissible dose of irradiation was reduced from 600 (1960) to 20 mSv/year for the employees and to 1 mSv/year for the population (1996) owing to the expansion of information on biological effect of radiation as well as to the working out of effective protective measures.

One can see that the limiting values, regulated by the rules, appear to approximate the level of a natural background radiation.

It is quite evident that the organization of protective measures from irradiation as well as of the work with sources of ionizing radiation are creating a certain economic burden for all kinds of work with different radiation sources and, in general, for the economic budget of each country. In the nearest future it could be expected that if the presence of a threshold of stochastic effects and of the radiation hormesis are going to get a convincing proof, then the dose limits might be reconsidered towards increasing. Though, probably, it is reasonable to place particular emphasis on the most vulnerable group of population – pregnant women and children – in order to secure for them the existing strict standards.

The accumulation of genetic defects can play an important role in considering the possible ill effects from irradiation. If up to now no genetic load is felt, this does not mean at all that it cannot reveal itself all of a sudden, so one has to pay attention and stay cautious. Analyzing the currently available information about radiation effect on the genetic level, one can believe that evidently dangerous is the double increase in genetic load. The irradiation dose that could lead to the double increase of mutations and hereditary abnormalities is called "the doubling dose." Geneticists came to the common conclusion that irradiation of all human society in the limits of the doubling dose has to be considered as a real hazard for humanity. At present, the doubling dose is estimated in the limits between 0.3 and 0.8 Gy. It has to be taken into consideration that all calculations of a genetic hazard are reasonable on the condition that the whole population is exposed to radiation. If only the small groups of the population are exposed, it would drastically reduce the likelihood of genetic disorders.

The existing standards and regulations are still subject to criticism. But development and implementation of ecologically safe regulations of ecological safety are faced with a number of difficulties related to some imperfections of both international (IAEA) and national normative juridical bases concerning this problem.

9.2 Chemical (Carcinogenic) Safety Regulation

The problem of chemical safety regulation is quite acute. The negative effect of chemical factors on the population, on social and industrial infrastructure, and on the ecological system is growing, in addition to the risk of emergency cases. Also in some countries the number of dangerous industries close to the breaking point or completely depleted technical and technological sources grows. The technogenic pollutions as a result of different industrial activities (including the elimination of chemical weapons) led to the accumulation of toxic industrial wastes, polluted territories, and waters.

The complex chemical compounds, especially in small concentrations, are very difficult to register. In Section 5.1.1, it was already pointed out how many chemical substances are used by mankind and how many of them are actually or potentially dangerous.

For all these perplexities, though, to regulate the use of chemicals and to normalize the toxicity indexes is necessary.

Because one of the main dangers of the health effects of low chemical doses is a danger of malignant neoplasm induction, the problem of carcinogenic security deserves greater consideration here.

In Section 5.4.1, the databases of information on many carcinogenic substances were presented.

Regulation of the use of possible carcinogens is based on the classification by hazard suggested by the founder of experimental oncology in the USSR L.M. Shabad. All chemicals were subdivided there into several categories. The first category consisted of chemicals with carcinogenicity proven by experiments on animals and by epidemiological observations. The second group includes chemicals that can cause a high percentage of tumors in laboratory animals of several species and in various ways of administration of carcinogenic chemicals. The third category includes chemicals that experimentally cause slow-growing tumors in only 20%–30% animals, and the fourth category includes agents with "equivocal" activity, which often produce contradictory results.

The modern classification only slightly differs from the one suggested by L.M. Shabad.

At present, The International Agency for Research on Cancer (IARC) suggests five groups for chemicals, mixtures, and impact factors.

Category 1: chemicals carcinogenic to humans; for these the absolute proof of tumor hazards in humans is not only the result of experiment on animals but also of convincing epidemiological data. The experts of IARC believe there exist more than 60 of such chemicals.

Category 2A: most probably (with a high degree of proof) carcinogenic to humans: 51 chemicals or factors.

Category 2B: probably ("possibly," with a lower degree of probability) carcinogenic to humans: 192 factors.

Category 3: not classified as carcinogenic to humans: 446 chemicals. Although there is no sufficient ground to refer them to the carcinogenic group, it is not quite possible to consider them nonhazardous.

Category 4: chemicals noncarcinogenic to humans. Valid proof must be given that they are not carcinogenic at all (so far the experts refer to this group only one chemical, that is, Caprolactam) [3,4].

Naturally, the numbers of factors above are being changed with the appearance of new data.

In the list above, the term "factor" is used, because it is not always possible clearly distinguish a specific chemical that represents carcinogenic hazard. Sometimes it is essential to indicate a certain kind of hazardous industry. It can be a production of amines' synthesis (fraught with bladder cancer), a processing of chromium products (mucosa of nasal cavity and lung mucosa cancer), a rubber manufacturing (lung cancer), or hematite mines (lung cancer).

In some instances, the complex mixture of different chemicals is going to be pointed out, because in such cases carcinogenesis is a result of a combined effect of various xenobiotics. Tobacco smoke, for example, can be considered such a mixture.

Analyzing the problem of carcinogenic hazard, one has to distinguish between the notions "carcinogenicity" and "carcinogenic hazard" of a chemical. Carcinogenicity defines the ability of a chemical to induce the development of malignant neoplasm and allows to compare chemicals affecting the body directly according to this principle. Carcinogenic hazard presupposes additional terms: the prevalence of a chemical, the possibility to contact with it, its stability in the environment, or in places of potential contacts, etc.

In order to regulate the content of carcinogens on work sites and in the environment, the same parameter as for noncarcinogenic substances is used, that is, maximum allowable concentration (MAC). But in this case, it can be used with one reservation. It is generally assumed for the noncarcinogenic chemicals that can potentially perform a toxic action, that they have no adverse effects on human health if their concentration is below a certain threshold level and, besides, its effect passes off immediately after its action stops. As to carcinogenic chemicals, there exists no evidence of existence of a concentration below which carcinogenic chemicals would get nonhazardous (the concept of nonthreshold action). Besides, the organism keeps in memory the effects already performed and goes on to accumulate as yet hidden genome damage (mutations). That's why there can be, strictly speaking, no limits for carcinogenic chemicals. According to the nonthreshold concept, not only is the development of MAC for carcinogens inadmissible but so is their presence in the environment altogether. But it is quite evident that it is not possible to completely exclude their presence in the environment. It is impossible because carcinogens appear not only as a result of a human technological activity but also owing to natural processes (as is well known, only the amount of benzopyrene, entering the Earth's atmosphere yearly with volcanic ashes, may range between 12 and 24 tons) as well as with cooking (e.g., while roasting) and in some other cases.

The problem can be solved with the use of the concept of acceptable risk. It is evident, that the lesser the effect either in intensity or in duration, the lesser is the morbidity risk. Based on this idea, the MAC of a toxicant or a virtually safe dose (VSD) of a carcinogen is administered.

The MAC of a carcinogen is such a concentration that while the employees contacting with such chemicals during their labor activity or the population that lives in the area where a chemical plant is located are exposed to it, the absence of carcinogenic effects is guaranteed with the specified high reliability. Naturally, it is only with the reliability that could be attained with the modern methods of epidemiological or experimental research.

The VSD is a toxicant dose that makes possible only a small increase in the number of neoplasms compared with the control group. Usually, this increase is considered as equal to 10^{-5} to 10^{-8} (one extra occurrence in 100 thousand to 100 million people, exposed to the action).

As the mechanisms of chemical and of radiation carcinogenesis are similar, the effects of small doses, namely, the existence of a threshold for stochastic effects and hormesis obtained with extensive experimental and observational material, are expected to be quite valid also for chemical carcinogenesis.

References

[1] Stabin MG. Radiation Protection and Dosimetry. An Introduction to Health Physics. Springer; 2007. p. 378. Ch. 7. The Basis for Regulation of Radiation Exposure, p. 105–131.(11 periods).

[2] Stabin MG. Fundamentals of Nuclear Medicine Dosimetry. Springer; 2008. p. 237 Ch.7. Regulatory Aspects of Dose Calculations, p. 201–220.

[3] IARC Monographs on the Evaluation of Carcinogenic Risk to Humans. Lyon, France, 2006 - http://monographs.iarc.fr/ENG/Preamble/CurrentPreamble.pdf.

[4] U.S. EPA Guidelines for Carcinogen Risk Assessment - http://www.epa.gov/raf/publications/pdfs/CANCER_GUIDELINES_FINAL_3-25-05.PDF.

Conclusion

While writing this book, I had in mind to take advantage of a rule emerging from the principles of perceptual psychology that I once found in an instruction for the papers introduced for some scientific conference. It read more or less laconic in English: "Tell them what you are going to tell them, then tell them, then tell them what you have told them." In Russian it actually looks longer.

So, what was I going to tell?

In the first place, my main attention was focused on the two sources of danger. The one is the most frightful and potentially very dangerous factor, namely, nuclear radiation. The other is the exposure of various chemical substances on the human body.

In the second place, considerable emphasis has been placed on the effect of these factors in smaller doses. The point is that even the small doses of radiation as well as those of chemical substances are in fact the factors that steadily and independent of one's wish affect all inhabitants of our environment.

The potentially colossal destructive power of nuclear energy, in general, as well as nuclear radiation, in particular, has been successfully kept in check for many years now. Thanks to that, the relative harm for the whole of humanity, in spite of the wide use of nuclear radiation in science, medicine, power engineering, and in various fields of industry and agriculture has been rather negligible, so that the existing fears seem rather exaggerated. But along with it the various chemical substances that exist in nature quite in hand, as it were, and also those that are made artificially by technological processes, are causing substantial harm.

The humanity of today, at the household level as well as at the level of the governmental and social solutions, significantly overestimate the fears of nuclear radiation and underestimates the dangers of different chemical substances for humans.

I'd like to conclude with a short story.

The Kemerovo University in Russia launched an expedition in 1996 to investigate the radiation situation to the north of the Kemerovo region in Siberia, mainly in the zone of underground nuclear explosion. Some experts were invited from Moscow and Saint-Petersburg to participate in it, with the author included. The underground nuclear explosion code-named "Quartz" took place on September 18, 1984. Officially, the aim of the explosion was the intratelluric seismic probe of the earth's crust in order to look for structures promising for mineral exploration. That was the 649th test after the initial explosion of August 29, 1949, of a nuclear charge and the 112th intratelluric nuclear explosion within the program of "The Use of Nuclear Explosion Technology in the Interests of National Economy." It was a so-called "peaceful" nuclear explosion. There were 124 such "peaceful" explosions, including 8 as group explosions, with an impressive number of charges; the total amount of 135 nuclear charges were exploded in wells, tunnels, and in a mine.

The most part of those underground nuclear explosions were set off at considerable depth so that fission products did not outcrop and that they remain buried in the cavity of explosion. Such explosions are called "camouflage." In some special cases, with so-called excavation explosions, which were set off for trenching, constructing canals, making reservoirs and cavities for storage of large amounts of wastes, to control emergency gas torches, the explosion cavity outcropped and the radioactive fission products got into the environment. These were set off during more than 10 such explosions. In some cases, the breakthrough of the radioactive substances onto the surface took place contrary to the initial plan because of incorrect consideration of the hydrogeological structure of the locality.

In about 100 km from the town of Kemerovo on a clearing in a pine and cedar grove, which shrank back before a nice brook called Tyshtym, the gusher was hit. Into this gusher 557 m deep was located a nuclear charge with an approximate energy release equivalent to 10 kt of trotyl. Note that 10 kt of trotyl equivalent equals an energy of 4.184×10^{13} J ~ 10^7 kW•hour, which corresponds to the annual energy consumption of a town with a population of up to 100,000. After the nuclear charge was located, the gusher was concreted.

The nuclear explosion in Kemerovo region proved to be clean. Our measurements showed that on the surface, in the vicinity of the wellhead, there was no radioactivity that could exceed the natural background levels.

The drilling rig was left as it was; it remained on its place above the epicenter of the explosion, looming like an alarming tower above the treetops, marking the explosion location from afar. Usually the drill operators, after completing the well dismantle and remove the rig. But the explosion in Kemerovo region was one of the last, and as other explosions were not planned in the vicinity, so the rig remained. To dismantle and to transport such a colossus might turn out more expensive than to construct a new one.

There was nothing threatening and troublesome in the rig itself, but just one look from the distance at it loomed above the trees could unsettle anyone, suggesting that that was the epicenter of a nuclear explosion. Thus the radiophobia revealed itself, which even the experienced nuclear physicists could not escape. Of course, since the first nuclear bombardments of Japan and, especially, owing to the hot confrontation years of "the cold war," the words "nuclear explosion" keeps on hiding a certain threat in itself.

The news of the expedition and of our arrival was spreading very fast about the neighborhood and the whole stream of visitors kept coming to meet us. They were the officials of local enterprises, farmers, and reporters from the Kemerovo TV. All were preoccupied with the question: how do matters stand with the radiation? Is it safe to drink milk of the cows that are grazing in the neighborhood meadows, is it safe to eat the meat of those cows, is it safe to gather and eat the mushrooms growing in the surrounding woods? What if the crops from the fields in the vicinity are radioactive? All people were rather anxious about the probable radiation poisoning of the local woods, fields, meadows, and rivers.

In his interview to a reporter of the Kemerovo TV, the author, standing in the clearing with the above-mentioned rig in the background, explained that all products from the neighborhood pastures, fields, and woods are quite safe for eating and drinking. And just on that occasion, answering the question of whether the local mushrooms are safe to eat, he pointed out: "A nonradioactive death-cup is immeasurably more poisonous than one kilogram of normal edible radioactive mushrooms."

The most essential thing is to distinguish between "fear" and "danger." Some phenomena may seem fearful but sometimes they prove to be not dangerous at all but, just the opposite, the really hazardous things may, at the same time, not seem so threatening. Thus, some people could be scared of black cats and of the number "13," which have nothing to do with real danger. And those same people could drink low-grade drinks or smoke low-quality tobacco without bothering that those things could be really dangerous.

So, I wish we all are well informed and be able to distinguish between the real and the seeming disaster.

Every now and then, the author had to deal with situations where the majority of quite normal, mentally safe people were definitely suffering from radiophobia.

The Greek word "phobos" means just "fear." And the posterior derivation of it, that is "phobia," has deepened its purport toward an "impulsive obsession with fear" or "unmotivated fear." Added as a suffix to other words, the word acquires the meaning of various "phobias." On the site www.phobia-fear-release.com, one can find a long list of several hundred different phobias, from a well known "claustrophobia" up to the quite exotic one, like "Paraskavedekatriaphobia," which means "a fear of Friday, 13." Only slightly shorter is the Wikipedia List of Phobias.

Radiophobia is the one from those long lists. And one of the main goals of this book is to help people overcome it or, at least, to reduce the worries connected with it.

Professor Eduard Aluker (*left*), head of the expedition researching the radiation situation in the area of underground nuclear explosion in the Kemerovo region, and the author (*right*), both armed with dosimetric equipment, standing at the explosion epicenter. Summer, 1996.

Subject Index

D

E

O

S

T

Printed in the United States
By Bookmasters